Frontiers in Mathematics

Advisory Editorial Board

More information about this series at http://www.springer.com/series/5388

José Antonio Ezquerro Fernández
Miguel Ángel Hernández Verón

Newton's Method: an Updated Approach of Kantorovich's Theory

 Birkhäuser

José Antonio Ezquerro Fernández
Department of Mathematics and Computation
University of La Rioja
La Rioja, Spain

Miguel Ángel Hernández Verón
Department of Mathematics and Computation
University of La Rioja
La Rioja, Spain

ISSN 1660-8046   ISSN 1660-8054   (electronic)
Frontiers in Mathematics
ISBN 978-3-319-55975-9   ISBN 978-3-319-55976-6   (eBook)
DOI 10.1007/978-3-319-55976-6

Library of Congress Control Number: 2017944725

Mathematics Subject Classification (2010): 47H99, 65J15, 65H10, 45G10, 34B15

Printed on acid-free paper

This book is published under the trade name Birkhäuser, www.birkhauser-science.com
The registered company is Springer International Publishing AG
The registered company address is: Gewerbestrasse 11, 6330 Cham, Switzerland

Dedicated to my daughter María,
for always being there for me. JAE.

Dedicated to my dear children
Ramón, Jorge, Miguel, Diego and Mercedes,
who are the engine of my life,
for their love and patience. MAH.

Preface

One of the most common problems in mathematics is the solution of a nonlinear equation $F(x) = 0$. This problem is not always easy to solve, since we cannot frequently obtain an exact solution to the previous equation, so that we usually look for a numerical approximation to a solution. In this case, we use approximation methods, which are generally iterative. The best known iteration to solve nonlinear equations is undoubtedly Newton's method:

$$x_{n+1} = x_n - [F'(x_n)]^{-1} F(x_n), \quad n \geq 0, \quad \text{with } x_0 \text{ given.}$$

The geometric interpretation of Newton's method is well known if F is a real function. In such a case, x_{n+1} is the point where the tangential line $y - F(x_n) = F'(x_n)(x - x_n)$ of the function $F(x)$ intersects the x-axis at the point $(x_n, F(x_n))$. The geometric interpretation of the complex Newton method, $F : \mathbb{C} \longrightarrow \mathbb{C}$, is given by Yau and Ben-Israel in [84]. In the general case, $F(x)$ is approximated at point x_n as $F(x) \approx L_n(x) = F(x_n) + F'(x_n)(x - x_n)$ and the zero of $L_n(x) = 0$ defines the new approximation x_{n+1}.

In spite of its simple principle [83], depending on the domain of F, Newton's method is applicable to various types of equations such as systems of nonlinear algebraic equations including matrix eigenvalue problems, differential equations, integral equations, etc., and even to random operator equations [8]. Hence, the method fascinates many researchers. However, as it is well known, a disadvantage of the method is that the initial approximation x_0 must be chosen sufficiently close to a true solution in order to guarantee its convergence. Finding a criterion for choosing x_0 is quite difficult. In this text we try to facilitate the choice of x_0 under conditions as general as possible.

The long history of Newton's method has already been well studied, see, e.g., N. Kollerstrom [57] or T. J. Ypma [85]. According to these articles, in [19], P. Deuflhard points out the facts seem to be agreed upon among the experts. The interested reader may find more historical details in the book by H. H. Goldstine [35].

In particular, as a summary, we rewrite some notes and remarks given by Ortega and Rheinboldt in [64], which is a good survey until 1970. Point-of-attraction results date back to the 19th century. For one-dimensional equations, the quadratic convergence of Newton's method was established by Cauchy (1829). For equations in $\mathbb{R}^n$, a point-of-attraction theorem was given by Runge (1899), who also stressed the quadratic convergence. Independently, a result for $n = 2$ was given by Blutel (1910). For convergence results which do not assume the existence of a solution, Fine (1916) appears to have been the first to prove the convergence of Newton's method in n dimensions, where the derivative $F'(x)$ is assumed to be invertible on some suitable ball. In the same year, Bennet (1916) formulated related results for operators on infinite dimensional spaces but for the proofs he referred to Fine. The article of Fine appears to have been overlooked, and twenty years later Ostrowski (1936) presented, independently, new convergence theorems and also discussed error estimates. Concurrently, Willers (1938)

also proved similar convergence conditions. Although both these authors observed that their
results extend immediately to the case of a general n, they themselves presented them only
for $n = 2$ and $n = 3$, respectively. Bussmann (1940), in an unpublished dissertation, proved
these results and some extensions for general n; Bussmann's theorems are quoted by Rehbock
(1942).

Later, the Russian mathematician L. V. Kantorovich gave his now famous convergence
results for Newton's method in Banach spaces. In 1948, Kantorovich published the seminal
paper [47], where he suggested an extension of Newton's method to functional spaces and
established a semilocal convergence result for Newton's method in a Banach space, which is
now called Kantorovich's theorem or, more specifically, *the Newton-Kantorovich theorem*, as
we will call from now on. The result was also included in the survey paper [48]. Further
developments of the method can be found in [49, 50, 51, 52, 53] and in the monographs [54, 55].

The main contribution of Kantorovich is the formulation of the problem in a general
setting, the spaces of Banach, that uses appropriate techniques of functional analysis. This
event cannot be overestimated, since Newton's method became a powerful tool in numerical
analysis as well as in pure mathematics. The approach of Kantorovich guarantees the application of Newton's method to solve a large variety of functional equations: nonlinear integral
equations, ordinary and partial differential equations, variational problems, etc. Various examples of such applications are presented in [54, 55]. For a list of relevant publications where
Newton's method is applied to different functional equations, see [64] and the references cited
there.

The Kantorovich result is a masterpiece not only by its sheer importance but by the
original and powerful proof technique. The results of Kantorovich and his school initi-
ated some very intensive research on the Newton and related methods. A great number
of variants and extensions of his results emerged in the literature. Basic results on Newton's
method and numerous references may be found in the books of Ostrowski [65] and Ortega and
Rheinboldt [64]. More recent bibliography is available in the books of Rheinboldt [74] and
Deulfhard [18], survey paper [85] and the special web site devoted to Newton's method [59].
A revision of the most important theoretical results on Newton's method concerning the
convergence properties, the error estimates, the numerical stability and the computational
complexity of the algorithm may be found in [33].

On the other hand, three types of studies can be done when we are interested in proving
the convergence of the sequence $\{x_n\}$ given by Newton's method: local, semilocal and global.
First, the local study of the convergence is based on demanding conditions to the solution x^*,
from certain conditions on the operator F, and provide the so-called ball of convergence ([16])
of the sequence $\{x_n\}$, that shows the accessibility to x^* from the initial approximation x_0
belonging to the ball. Second, the semilocal study of the convergence is based on demanding
conditions to the initial approximation x_0, from certain conditions on the operator F, and
provide the so-called domain of parameters ([30]) corresponding to the conditions required
to the initial approximation that guarantee the convergence of the sequence $\{x_n\}$ to the
solution x^*. Third, the global study of the convergence guarantees, from certain conditions
on the operator F, the convergence of the sequence $\{x_n\}$ to the solution x^* in a domain and
independently of the initial approximation x_0. The three studies demand conditions on the
operator F. However, requirement of conditions to the solution, to the initial approximation,
or to none of these, determines the different types of studies.

The local study of the convergence has the disadvantage of being able to guarantee that the solution, that is unknown, can satisfy certain conditions. In general, the global study of the convergence is very specific as regards the type of operators to consider, as a consequence of absence of conditions on the initial approximation and on the solution. There is a plethora of studies on the weakness and/or extension of the hypothesis made on the underlying operators. In this textbook, we focus our attention on the analysis of the semilocal convergence of Newton's method.

This textbook is written for researchers interested in the theory of Newton's method in Banach spaces. Each chapter contains several theoretical results and interesting applications in the solution of nonlinear integral and differential equations.

Chapter 1 presents an analysis of Kantorovich's theory for Newton's method where the original theory is given along with the best-known variant, which is due to Ortega and uses the method of majorizing sequences. In addition, we include a new approach by introducing a new concept of majorant function which is different from that defined by Kantorovich. In all the results presented, we suppose that the second derivative of the operator involved is bounded in norm in the domain where the operator is defined.

In Chapter 2, we analyse the semilocal convergence of Newton's method under different modifications of the condition on the second derivative of the operator involved. We begin by presenting the result of Huang where the second derivative of the operator involved is Lipschitz continuous in the domain where the operator is defined. We pay attention to the proof given by Huang and see that the condition on the second derivative can be relaxed to a center Lipschitz condition to establish the semilocal convergence of Newton's method. This observation leads us to propose milder conditions on the second derivative of the operator involved. So, we first prove the semilocal convergence of Newton's method under a center ω-Lipschitz condition for the second derivative of the operator involved. Next, the condition of the second derivative of the operator required by Kantorovich is generalized to a condition where the second derivative is ω-bounded in the domain where the operator is defined. This generalization is very interesting because it avoids having to look for a domain where the second derivative of the operator is bounded and contains a solution of the equation to solve, which is an important problem that presents Kantorovich's theory.

The ideas developed in Chapter 2 are extended to higher order derivatives of the operator involved in Chapter 3, where new starting points for Newton's method are located despite the new conditions imposed to the operator. Three sections are included where the semilocal convergence of Newton's method is analysed for polynomial operators, operators with ω-bounded k-th-derivative and operators with ω-Lipschitz k-th-derivative.

Chapter 4 examines the typical situation in which conditions on the first derivative of the operator are only required, instead of conditions on higher order derivatives, as a consequence of the fact that this is the only derivative of the operator involved appearing in the algorithm of the method. We analyse in detail Ortega's variant of the original Kantorovich's result from the method of majorizing sequences and pay special attention to the fact that, if we want to relax the conditions imposed to the first derivative of the operator, we need to apply another distinct technique to that of the method of majorizing sequences presented in Chapter 1 to guarantee the semilocal convergence of Newton's method. In particular, we propose a system of recurrence relations. Techniques based on recurrence relations were already used by Kantorovich in his first proofs of the semilocal convergence of Newton's method. We then analyse two cases: the cases in which the first derivative of the operator involved is ω-Lipschitz continuous and center ω-Lipschitz continuous in the domain where the operator involved is defined.

We have included a numerical example in each situation analysed theoretically to so justify its analysis. In particular, in Chapter 1, we present two types of problems that are used throughout the following chapters: nonlinear integral equations of Hammerstein type and nonlinear boundary value problems. We analyse these problems some times on a continuous form and others in a discrete way.

We emphasize the fact that we have tried to facilitate the reading of the textbook from a detailed development of the proofs of the results given. We also want to point out that the textbook presents some of our investigations into Newton's method which we have carried out over many years, including an abundant and specialized bibliography.

We ended up saying that the main aim of this textbook is to develop, expand and update the theory introduced by Kantorovich for Newton's method under different conditions on the operator involved, but always with the clear objective to improve the applicability of the method based on the location of new starting points.

Logroño, La Rioja   J. A. Ezquerro
June 2016   M. A. Hernández-Verón

Contents

Preface   vii

1 The classic theory of Kantorovich   1
1.1 The Newton-Kantorovich theorem   1
    1.1.1 Recurrence relations of Kantorovich   2
    1.1.2 The majorant principle of Kantorovich   7
    1.1.3 The "method of majorizing sequences" 17
    1.1.4 New approach: a "majorant function" from an initial value problem . . 24
1.2 Applications 28
    1.2.1 Integral equations 29
    1.2.2 Boundary value problems 32

2 Convergence conditions on the second derivative of the operator   39
2.1 Operators with Lipschitz-type second-derivative 40
    2.1.1 Operators with Lipschitz second-derivative 40
    2.1.2 Operators with Hölder second-derivative 46
    2.1.3 Operators with ω-Lipschitz second-derivative 47
        2.1.3.1 Existence of a majorizing sequence 48
        2.1.3.2 Existence of a majorant function 54
        2.1.3.3 Semilocal convergence 55
        2.1.3.4 Uniqueness of solution and order of convergence 55
        2.1.3.5 Applications 57
        2.1.3.6 A little more: three particular cases 63
2.2 Operators with ω-bounded second-derivative 70
    2.2.1 Existence of a majorant function 72
    2.2.2 Existence of a majorizing sequence 73
    2.2.3 Semilocal convergence 75
    2.2.4 Uniqueness of solution and order of convergence 76
    2.2.5 Applications 78

3 Convergence conditions on the k-th derivative of the operator   83
3.1 Polynomial type equations 83
    3.1.1 Semilocal convergence 85
    3.1.2 Uniqueness of solution and order of convergence 89
    3.1.3 Applications 91
3.2 Operators with ω-bounded k-th-derivative 95
    3.2.1 Existence of a majorizing sequence 96

3.2.2   Existence of a majorant function 102
3.2.3   Semilocal convergence 103
3.2.4   Uniqueness of solution and order of convergence 104
3.2.5   Application 105
3.3   Operators with ω-Lipschitz k-th-derivative 109
3.3.1   Existence of a majorizing sequence 110
3.3.2   Existence of a majorant function 116
3.3.3   Semilocal convergence 118
3.3.4   Uniqueness of solution and order of convergence 118
3.3.5   Application 120
3.3.6   Relaxing convergence conditions 122

4   Convergence conditions on the first derivative of the operator   127
4.1   Operators with Lipschitz first-derivative 128
4.1.1   Existence of a majorizing sequence 128
4.1.2   Existence of a majorant function 132
4.1.3   Semilocal convergence 132
4.2   Operators with ω-Lipschitz first-derivative 134
4.2.1   Semilocal convergence using recurrence relations 135
4.2.1.1   Recurrence relations 135
4.2.1.2   Analysis of the scalar sequences 136
4.2.1.3   Semilocal convergence 137
4.2.2   Uniqueness of solution and order of convergence 139
4.2.3   Application 143
4.2.4   Particular cases 147
4.2.4.1   Operators with Lipschitz first-derivative 148
4.2.4.2   Operators with Hölder first-derivative 149
4.3   Operators with center ω-Lipschitz first-derivative 150
4.3.1   Semilocal convergence 151
4.3.2   Particular case: operators with center Lipschitz first-derivative 156
4.3.3   Application 157

Bibliography   161

Chapter 1

The classic theory of Kantorovich

According to Polyak [66], Kantorovich proved in 1939 the semilocal convergence of Newton's method [46] on the basis of the contraction mapping principle of Banach, and later improved to semilocal quadratic convergence in 1948/49 (*the Newton-Kantorovich theorem*) [47, 49]. Also in 1949, Mysovskikh [61] gave a much simpler independent proof of semilocal quadratic convergence under slightly different theoretical assumptions, which are exploited in modern Newton algorithms, see [18].

Kantorovich gave two basically different proofs of the Newton-Kantorovich theorem using recurrence relations or majorant functions. The original proof given by Kantorovich uses recurrence relations [47]. A nice treatment of this theorem using recurrence relations can be found in [72]. In [50] Kantorovich gave a proof based on the concept of a majorant function.

An important feature of the Newton-Kantorovich theorem, or related results, is that it does not assume the existence of a solution, so that the theorem is not only a convergence result for Newton's method, but simultaneously a theorem of existence of solution for nonlinear equations in Banach spaces. In addition, the theoretical significance of Newton's method can be used to draw conclusions about the existence and uniqueness of a solution and about the region in which it is located, without finding the solution itself and this is sometimes more important than the actual knowledge of the solution (see [55]). The results of this section follows Kantorovich's paper [53].

After Kantorovich establishes the Newton-Kantorovich theorem, a large number of results has been published concerning convergence and error bounds for Newton's method under the assumptions of the Newton-Kantorovich theorem or under closely related ones. Among later convergence theorems, the ones due to Ortega and Rheinboldt [64] are worth mentioning. Also, there exist numerous versions of the Newton-Kantorovich theorem that differ in assumptions and results, and it would be impossible to list all relevant publications here. We then only mention some versions that may find in [11, 54, 55, 58, 62, 65].

Throughout the textbook we denote $\overline{B(x, \varrho)} = \{y \in X; \|y - x\| \leq \varrho\}$ and $B(x, \varrho) = \{y \in X; \|y - x\| < \varrho\}$.

1.1 The Newton-Kantorovich theorem

Newton's method has the form

$$x_{n+1} = N_F(x_n) = x_n - [F'(x_n)]^{-1}F(x_n), \quad n \geq 0, \quad \text{with } x_0 \text{ given,} \qquad (1.1)$$

© Springer International Publishing AG 2017
J.A. Ezquerro Fernández, M.Á. Hernández Verón, *Newton's Method: an Updated Approach of Kantorovich's Theory*, Frontiers in Mathematics, DOI 10.1007/978-3-319-55976-6_1

for the solution of the nonlinear equation $F(x) = 0$, where $F : X \longrightarrow Y$, X and Y are Banach spaces and F is twice continuously Fréchet differentiable. Below, we present some results given by Kantorovich on the convergence of Newton's method.

1.1.1   Recurrence relations of Kantorovich

Kantorovich established a semilocal convergence theorem for Newton's method in a Banach space in 1948 under the following conditions for the operator F and the starting point x_0 [47]:

(W1) For $x_0 \in X$, there exists $\Gamma_0 = [F'(x_0)]^{-1} \in \mathcal{L}(Y, X)$, where $\mathcal{L}(Y, X)$ is the set of bounded linear operators from Y to X, such that $\|\Gamma_0\| \leq \beta$,

(W2) $\|\Gamma_0 F(x_0)\| \leq \eta$,

(W3) there exists $R > 0$ such that $\|F''(x)\| \leq M$, for $x \in B(x_0, R)$,

(W4) $h = M\beta\eta \leq \frac{1}{2}$.

Theorem 1.1. (The Newton-Kantorovich theorem) *Let $F : \Omega \subseteq X \longrightarrow Y$ be a twice continuously Fréchet differentiable operator defined on a non-empty open convex domain Ω of a Banach space X with values in a Banach space Y. Suppose that conditions (W1)-(W2)-(W3)-(W4) are satisfied and $B(x_0, \rho^*) \subset B(x_0, R)$ with $\rho^* = \frac{1-\sqrt{1-2h}}{h}\eta$. Then, Newton's sequence defined in (1.1) and starting at x_0 converges to a solution x^* of the equation $F(x) = 0$ and the solution x^* and the iterates x_n belong to $\overline{B(x_0, \rho^*)}$, for all $n \geq 0$ Moreover, if $h < \frac{1}{2}$, the solution x^* is unique in $B(x_0, \rho^{**}) \cap B(x_0, R)$, where $\rho^{**} = \frac{1+\sqrt{1-2h}}{h}\eta$, and, if $h = \frac{1}{2}$, x^* is unique in $\overline{B(x_0, \rho^*)}$. Furthermore, we have the following error estimates:*

$$\|x^* - x_n\| \leq \frac{1}{2^{n-1}}(2h)^{2^n-1}\eta, \quad n = 0, 1, 2, \ldots \tag{1.2}$$

Proof. First of all, we define $\beta_0 = \beta$, $\eta_0 = \eta$ and $h_0 = h$. After that, we observe that x_1 is well-defined, since the operator $\Gamma_0 = [F'(x_0)]^{-1}$ exists by the hypotheses. Moreover, $\|x_1 - x_0\| \leq \eta_0$, so that $x_1 \in B(x_0, \rho^*)$.

Next, taking into account that

$$\|I - \Gamma_0 F'(x_1)\| \leq \|\Gamma_0\|\|F'(x_0) - F'(x_1)\|$$

$$\leq \beta_0 \left\| \int_0^1 F''(x_0 + \tau(x_1 - x_0))\,d\tau(x_1 - x_0) \right\|$$

$$\leq M\beta_0\|x_1 - x_0\|$$

$$\leq M\beta_0\eta_0$$

$$= h_0$$

$$< 1,$$

it follows, by the Banach lemma on invertible operators [72], that the operator $\Gamma_1 = [F'(x_1)]^{-1}$ exists and

$$\|\Gamma_1\| \leq \frac{\beta_0}{1 - h_0} = \beta_1.$$

Hence, x_2 is well-defined.

In addition, as

$$F(x_1) = F(x_0) + F'(x_0)(x_1 - x_0) + \int_{x_0}^{x_1} F''(z)(x_1 - z)\, dz$$

$$= \int_{x_0}^{x_1} F''(z)(x_1 - z)\, dz$$

$$= \int_0^1 F''(x_0 + \tau(x_1 - x_0))(x_1 - x_0)^2 (1 - \tau)\, d\tau,$$

we have

$$\|F(x_1)\| \leq \int_0^1 \|F''(x_0 + \tau(x_1 - x_0))\|\,(1 - \tau)\, d\tau \|x_1 - x_0\|^2$$

$$\leq \frac{M}{2}\|x_1 - x_0\|^2$$

$$\leq \frac{M}{2}\eta_0^2$$

and

$$\|x_2 - x_1\| = \|\Gamma_1 F(x_1)\| \leq \|\Gamma_1\|\|F(x_1)\| \leq \frac{\beta_0}{1 - h_0}\frac{M}{2}\eta_0^2 = \frac{h_0\eta_0}{2(1 - h_0)} = \eta_1 \leq h_0\eta_0.$$

Moreover,

$$h_1 = M\beta_1\eta_1 = \frac{h_0^2}{2(1 - h_0)^2} \leq 2h_0^2 \leq \frac{1}{2}.$$

Therefore, the hypotheses of the Newton-Kantorovich theorem are also satisfied when we substitute β_1, η_1 and h_1 for β_0, η_0 and h_0, respectively. This allows us to continue the successive determination of the elements x_n and the numbers connected with them, β_n, η_n and h_n, so that if we assume

$$\|\Gamma_n\| \leq \beta_n = \frac{\beta_{n-1}}{1 - h_{n-1}}, \tag{1.3}$$

$$\|x_{n+1} - x_n\| \leq \frac{M}{2}\beta_n\eta_{n-1}^2 = \frac{h_{n-1}\eta_{n-1}}{2(1 - h_{n-1})} = \eta_n \leq h_{n-1}\eta_{n-1} \leq \frac{1}{2^n}(2h_0)^{2^n - 1}\eta_0, \tag{1.4}$$

$$h_n = M\beta_n\eta_n = \frac{h_{n-1}^2}{2(1 - h_{n-1})^2} \leq 2h_{n-1}^2 \leq \frac{1}{2}, \tag{1.5}$$

where the operator $\Gamma_n = [F'(x_n)]^{-1}$ exists, it follows the following.

As

$$\|I - \Gamma_n F'(x_{n+1})\| \leq \|\Gamma_n\|\|F'(x_n) - F'(x_{n+1})\|$$

$$\leq \beta_n \left\| \int_0^1 F''(x_n + \tau(x_{n+1} - x_n))\, d\tau (x_{n+1} - x_n) \right\|$$

$$\leq M\beta_n\|x_{n+1} - x_n\|$$

$$\leq M\beta_n\eta_n$$

$$= h_n$$

$$< 1,$$

we have, by the Banach lemma on invertible operators, that the operator Γ_{n+1} exists and

$$\|\Gamma_{n+1}\| \leq \frac{\beta_n}{1 - h_n} = \beta_{n+1}.$$

Hence, x_{n+2} is well-defined.

Besides, as

$$F(x_{n+1}) = F(x_n) + F'(x_n)(x_{n+1} - x_n) + \int_{x_n}^{x_{n+1}} F''(z)(x_{n+1} - z)\, dz$$

$$= \int_{x_n}^{x_{n+1}} F''(z)(x_{n+1} - z)\, dz$$

$$= \int_0^1 F''(x_n + \tau(x_{n+1} - x_n))(x_{n+1} - x_n)^2 (1 - \tau)\, d\tau$$

we have

$$\|F(x_{n+1})\| \leq \int_0^1 \|F''(x_n + \tau(x_{n+1} - x_n))\|(1 - \tau)\, d\tau \|x_{n+1} - x_n\|^2$$

$$\leq \frac{M}{2}\|x_{n+1} - x_n\|^2$$

$$\leq \frac{M}{2}\eta_n^2$$

and

$$\|x_{n+2} - x_{n+1}\| = \|\Gamma_{n+1}F(x_{n+1})\| \leq \|\Gamma_{n+1}\|\|F(x_{n+1})\| \leq \frac{M}{2}\beta_{n+1}\eta_n^2 = \frac{h_n\eta_n}{2(1 - h_n)} = \eta_{n+1}.$$

Moreover, since $h_n \leq \frac{1}{2}$, we also have

$$\eta_{n+1} \leq h_n\eta_n \leq \cdots \leq h_n h_{n-1} \cdots h_0\eta_0 \leq \frac{1}{2^{n+1}}(2h_0)^{2^{n+1}-1}\eta_0$$

and

$$h_{n+1} = M\beta_{n+1}\eta_{n+1} = \frac{h_n^2}{2(1 - h_n)^2} \leq 2h_n^2 \leq \cdots \leq \frac{1}{2}(2h_0)^{2^{n+1}} \leq \frac{1}{2}.$$

As a consequence, (1.3), (1.4) and (1.5) are true for all positive integers n by mathematical induction.

On the other hand, if we note the identity

$$\eta_n\varphi(h_n) - \eta_{n+1}\varphi(h_{n+1}) = \eta_n,$$

where $\varphi(t) = \frac{1 - \sqrt{1 - 2t}}{t}$, which is verificable directly, since

$$\eta_{n+1}\varphi(h_{n+1}) = \eta_{n+1}\frac{1 - \sqrt{1 - 2h_{n+1}}}{h_{n+1}} = \eta_n\frac{1 - h_n - \sqrt{1 - 2h_n}}{h_n} = \eta_n\varphi(h_n) - \eta_n,$$

it follows, by (1.5), for $m \geq 1$ and $n \geq 1$, that

$$\|x_{n+m} - x_n\| \leq \|x_{n+m} - x_{n+m-1}\| + \|x_{n+m-1} - x_{n+m-2}\| + \cdots + \|x_{n+1} - x_n\|$$

$$\leq \eta_{n+m-1} + \eta_{n+m-2} + \cdots + \eta_n$$

$$= \eta_n\varphi(h_n) - \eta_{n+m}\varphi(h_{n+m})$$

$$\leq \eta_n\varphi(h_n) \tag{1.6}$$

$$\leq 2\eta_n$$

$$\leq \frac{1}{2^{n-1}}(2h_0)^{2^n-1}\eta_0, \tag{1.7}$$

so that $\{x_n\}$ is a Cauchy sequence and then convergent. In addition, passing to the limit in (1.7) as $m \to +\infty$, we obtain (1.2).

If we now do $n = 0$ in (1.6), then

$$\|x_m - x_0\| \leq \eta_0 \varphi(h_0) = \rho^*,$$

so that $x_m \in \overline{B(x_0, \rho^*)}$, for all $m \in \mathbb{N}$. Moreover, $\lim_n x_n = x^* \in \overline{B(x_0, \rho^*)}$. Furthermore, x^* is a solution of $F(x) = 0$, since $\|\Gamma_n F(x_n)\| = \|x_{n+1} - x_n\| \to 0$, when $n \to +\infty$, $\|F(x_n)\| \leq \|F'(x_n)\| \|x_{n+1} - x_n\|$, the sequence $\{\|F'(x_n)\|\}$ is bounded, since

$$\begin{aligned}
\|F'(x_n)\| &\leq \|F'(x_0)\| + \|F'(x_n) - F'(x_0)\| \\
&\leq \|F'(x_0)\| + M\|x_n - x_0\| \\
&\leq \|F'(x_0)\| + M\eta_0 f(h_0),
\end{aligned}$$

and $\|F(x_n)\| \to 0$ as $n \to +\infty$. Therefore, by the continuity of F in $\overline{B(x_0, \rho^*)}$, we obtain $F(x^*) = 0$.

Finally, we prove the uniqueness of the solution x^*. We first analyse the case $h < \frac{1}{2}$. Suppose that there exists a solution $y^* \in B(x_0, \rho^{**}) \cap B(x_0, R)$ of $F(x) = 0$ and such that $y^* \neq x^*$. Then, we have $\|y^* - x_0\| \leq \theta \rho^{**} = \theta \eta_0 \psi(h_0)$, where $\theta \in (0, 1)$ and $\psi(t) = \frac{1 + \sqrt{1 - 2t}}{t}$. Next, we suppose $\|y^* - x_j\| \leq \theta^{2^j} \eta_j \, \psi(h_j)$, for $j = 0, 1 \ldots, n$, and prove $\|y^* - x_{n+1}\| \leq \theta^{2^{n+1}} \eta_{n+1} \psi(h_{n+1})$.

Indeed, from $F(y^*) = 0$ and $x_{n+1} = x_n - \Gamma_n F(x_n)$, it follows

$$\begin{aligned}
y^* - x_{n+1} &= y^* - x_n + \Gamma_n F(x_n) \\
&= -\Gamma_n \left(F(y^*) - F(x_n) - F'(x_n)(y^* - x_n) \right) \\
&= -\Gamma_n \int_{x_n}^{y^*} F''(z)(y^* - z) \, dz \\
&= -\Gamma_n \int_0^1 F''(x_n + \tau(y^* - x_n))(y^* - x_n)^2 (1 - \tau) \, d\tau
\end{aligned} \tag{1.8}$$

and

$$\begin{aligned}
\|y^* - x_{n+1}\| &\leq \frac{M}{2} \|\Gamma_n\| \|y^* - x_n\|^2 \\
&\leq \frac{M}{2} \beta_n \left(\theta^{2^n} \eta_n \psi(h_n) \right)^2 \\
&= \theta^{2^{n+1}} \eta_{n+1} \psi(h_{n+1}).
\end{aligned}$$

Then, by mathematical induction, we have proved that $\|y^* - x_j\| \leq \theta^{2^j} \eta_j \psi(h_j)$ are true for all positive integers j.

Now, as

$$\eta_n \psi(h_n) = \eta_n \frac{1 + \sqrt{1 - 2h_n}}{h_n} \leq \frac{2\eta_n}{h_n} = \frac{2}{M\beta_n}$$

and $\beta_0 < \beta_n$, for $n \geq 0$, we have

$$\|y^* - x_n\| \leq \theta^{2^n} \frac{2}{M\beta_n} < \theta^{2^n} \frac{2}{M\beta_0}$$

and therefore $\|y^* - x_n\| \to 0$ as $n \to +\infty$, so that $y^* = x^*$, since $x^* = \lim_n x_n$.

For the case $h = \frac{1}{2}$, we suppose that y^* is a solution of $F(x) = 0$ in $\overline{B}(x_0, \rho^*)$ and such that $y^* \neq x^*$. Moreover, $\|y^* - x_0\| \leq \rho^* = 2\eta_0$. We now suppose $\|y^* - x_j\| \leq \frac{\eta_0}{2^{j-1}}$, for $j = 0, 1\ldots, n$, and prove $\|y^* - x_{n+1}\| \leq \frac{\eta_0}{2^n}$. Indeed, from (1.8), it follows

$$\|y^* - x_{n+1}\| \leq \frac{M}{2}\|\Gamma_n\|\|y^* - x_n\|^2 \leq \frac{M}{2}\beta_n(2\eta_n)^2 = 2h_n\eta_n \leq \eta_n \leq \frac{\eta_0}{2^n}.$$

Then, by mathematical induction on j, we conclude that $\|y^* - x_j\| \leq \frac{\eta_0}{2^{j-1}}$ are true for all positive integers j. As a consequence, $\|y^* - x_n\| \to 0$ as $n \to +\infty$ and therefore $y^* = x^*$. ∎

Note that condition (W4) of the Newton-Kantorovich theorem, which is often called *the Kantorovich condition*, is critical, since it means that, at the initial approximation x_0, the value $\|F(x_0)\|$ should be small enough, that is, x_0 should be close to a solution.

According to Galantai [33], if conditions (W1)-(W2)-(W3)-(W4) of the Newton-Kantorovich theorem are satisfied, then not only the Newton sequence $\{x_n\}$ exists and converges to a solution x^* but $[F'(x^*)]^{-1}$ exists in this case. Rall proves in [71] that the existence of $[F'(x^*)]^{-1}$ conversely guarantees that the hypotheses of the Newton-Kantorovich theorem with $h < \frac{1}{2}$ are satisfied at each point of an open ball with center x^*.

Notice that the Newton iterates x_n are invariant under any affine transformation $F \longrightarrow G = AF$, where A denotes any bounded and bijective linear mapping from Y to any Banach space Z [33]. This property is easily verified, since $[G'(x)]^{-1}G(x) = [F'(x)]^{-1}A^{-1}AF(x) = [F'(x)]^{-1}F(x)$. The affine invariance property is clearly reflected in the Newton-Kantorovich theorem. For other affine invariant theorems, we may see Deuflhard and Heindl [17].

Remark 1.2. The speed of convergence of an iterative method is usually measured by the order of convergence of the method. The first definition of order of convergence was given in 1870 by Schröder [76], but a very commonly measure of speed of convergence in Banach spaces is the R-order of convergence [69], which is defined as follows:

Let $\{x_n\}$ a sequence of points of a Banach space X converging to a point $x^* \in X$ and let $\sigma \geq 1$ and

$$e_n(\sigma) = \begin{cases} n & \text{if } \sigma = 1, \\ \sigma^n & \text{if } \sigma > 1, \end{cases} \quad n \geq 0.$$

(a) We say that σ is an R-order of convergence of the sequence $\{x_n\}$ if there are two constants $b \in (0, 1)$ and $B \in (0, +\infty)$ such that

$$\|x_n - x^*\| \leq Bb^{e_n(\sigma)}.$$

(b) We say that σ is the exact R-order of convergence of the sequence $\{x_n\}$ if there are four constants $a, b \in (0, 1)$ and $A, B \in (0, +\infty)$ such that

$$Aa^{e_n(\sigma)} \leq \|x_n - x^*\| \leq Bb^{e_n(\sigma)}, \quad n \geq 0.$$

In general, check double inequalities of (b) is complicated, so that normally only seek upper inequalities as (a). Therefore, if we find an R-order of convergence σ of sequence $\{x_n\}$, we then say that sequence $\{x_n\}$ has order of convergence at least σ. So, according to this, estimates (1.2) guarantee that Newton's method has R-order of convergence ([69]) at least two if $h < \frac{1}{2}$ and at least one if $h = \frac{1}{2}$.

1.1.2 The majorant principle of Kantorovich

Again according to Polyak [66], three years later (1951), Kantorovich gave another proof of the Newton-Kantorovich theorem and its versions, based on the so-called "majorant principle" [50, 51]. The idea is to compare iteration (1.1) with a scalar iteration that, in a sense, majorizes (1.1) and has the convergence properties. This approach provides more flexibility and the original proof is a particular case corresponding to a quadratic majorant, as we can see in [55].

In particular, Kantorovich proves the semilocal convergence of Newton's method from the semilocal convergence of the method of successive approximations. For this, Kantorovich considers the equation

$$x = \mathfrak{S}(x), \tag{1.9}$$

where $\mathfrak{S}$ is an operator defined on the ball $B(x_0, R)$ of some Banach space X with values in X, $x_0 \in X$, the scalar equation

$$t = \mathfrak{g}(t), \tag{1.10}$$

where the scalar function $\mathfrak{g}$ is defined on the interval $[t_0, t']$ with $t_0 \in \mathbb{R}$ and $t' = t_0 + r < t_0 + R$ and says that equation (1.10) (or the function $\mathfrak{g}$) *majorizes* equation (1.9) (or the operator $\mathfrak{S}$) if

$$\|\mathfrak{S}(x_0) - x_0\| \le \mathfrak{g}(t_0) - t_0, \tag{1.11}$$

$$\|\mathfrak{S}'(x)\| \le \mathfrak{g}'(t) \quad \text{when} \quad \|x - x_0\| \le t - t_0, \tag{1.12}$$

for $x \in B(x_0, R)$ and $t \in [t_0, t']$. Using this asumption, the convergence of the iterative method $x_{n+1} = \mathfrak{S}(x_n)$, $n \ge 0$, in X is deduced from that of the iterative method $t_{n+1} = \mathfrak{g}(t_n)$, $n \ge 0$, on the real line, and concludes that

$$\|x_{n+1} - x_n\| \le t_{n+1} - t_n, \quad n \ge 0, \tag{1.13}$$

so that the sequence $\{x_n\}$ is convergent if the scalar sequence $\{t_n\}$ is. Also, Kantorovich observes that Newton's method for equation $F(x) = 0$ coincides with the method of successive approximations applied to the equation $x = \mathfrak{S}(x) = x - [F'(x)]^{-1}F(x)$, which obviously is equivalent to $F(x) = 0$.

This ingenious procedure devised by Kantorovich to prove the convergence of a sequence in a Banach space from that of a scalar sequence has the disadvantage that the conditions required to the scalar function majorizes the operator of the Banach space are restrictive and, in the case of considering other iterative methods, these conditions are difficult to satisfy, basically condition (1.12). Even if we want to apply this technique of Kantorovich directly to the operator $\mathfrak{S}(x) = x - [F'(x)]^{-1}F(x)$, quite a few difficulties already appear (see [38]). For this, in this textbook, we present Kantorovich's theory to study the semilocal convergence of Newton's method under different conditions from another point of view. In particular, we base our development on a concept that is fundamental in Kantorovich's theory, which is the fact that a scalar sequence $\{t_n\}$ verifies (1.13). Now, in the following, we develop the majorant principle of Kantorovich.

Theorem 1.3. *Suppose the operator $\mathfrak{S}$ has a continuous first Fréchet derivative in a closed ball $\overline{B(x_0, r)}$ and the scalar function $\mathfrak{g}$ is derivable in the interval $[t_0, t']$ with $t' - t_0 = r$. If equation (1.10) majorizes equation (1.9) and equation (1.10) has a solution in $[t_0, t']$, then*

*(1.9) has a solution x^* and the sequence $\{x_n\}$, given by the method of successive approxima-
tions $x_{n+1} = \mathfrak{S}(x_n)$, $n \geq 0$, starting at x_0, converges to x^*. Moreover, $\|x^* - x_0\| \leq t^* - t_0$,
where t^* is the samllest positive solution of (1.10) in $[t_0, t']$.*

Proof. Note that $\mathfrak{S}$ is defined in all the points of $\overline{B(x_0, r)}$, since $\overline{B(x_0, r)} \subset B(x_0, R)$. If
t^* is the samllest positive solution of equation (1.10) in $[t_0, t']$, then $t_0 \leq t^*$. In addition, if
we suppose $t_i \leq t^*$, with $i = 0, 1, \ldots, n$, then $t_{n+1} = \mathfrak{g}(t_n) \leq \mathfrak{g}(t^*) = t^*$, since the function
$\mathfrak{g}$ is nondecreasing (as we see from condition (1.12)). As a consequence, by mathematical
induction on n, it follows easily that $t_n \leq t^*$, for all $n \geq 0$.

Next, from condition (1.11), we have $0 \leq \mathfrak{g}(t_0) - t_0 = t_1 - t_0$, so that $t_1 \geq t_0$. In addition,
from a mathematical induction on n, it follows that $t_n \leq t_{n+1}$, $n \geq 0$, since the function $\mathfrak{g}$ is
nondecreasing.

Therefore, there exists $v = \lim_n t_n$ such that $v \leq t^*$. Besides, as $\mathfrak{g}$ is a continuous function,
it follows

$$v = \lim_{n \to +\infty} t_{n+1} = \lim_{n \to +\infty} \mathfrak{g}(t_n) = \mathfrak{g}\left(\lim_{n \to +\infty} t_n\right) = \mathfrak{g}(v),$$

and then $v = t^*$ is solution of equation (1.10). As a consequence, the sequence $\{t_n\}$ converges
to the smallest positive solution t^* of (1.10).

After that, we see that the sequence $\{x_n\}$ is well defined; namely, $x_n \in \overline{B(x_0, r)}$ for $n \in \mathbb{N}$.
From condition (1.11), we obtain

$$\|x_1 - x_0\| = \|\mathfrak{S}(x_0) - x_0\| \leq \mathfrak{g}(t_0) - t_0 = t_1 - t_0 \leq t' - t_0 = r,$$

so that $x_1 \in \overline{B(x_0, r)}$.

Assume that it has already been proved that $x_1, x_2, \ldots, x_n \in \overline{B(x_0, r)}$ and

$$\|x_{i+1} - x_i\| \leq t_{i+1} - t_i, \quad \text{for} \quad i = 0, 1, \ldots, n - 1, \tag{1.14}$$

and prove by induction that $x_{n+1} \in \overline{B(x_0, r)}$ and (1.14) are true for all $i \geq 0$. Indeed, from

$$x_{n+1} - x_n = \mathfrak{S}(x_n) - \mathfrak{S}(x_{n-1}) = \int_{x_{n-1}}^{x_n} \mathfrak{S}'(z)\,dz = \int_0^1 \mathfrak{S}'\left(x_{n-1} + \tau(x_n - x_{n-1})\right)(x_n - x_{n-1})\,d\tau,$$

taking into account that

$$\begin{aligned}
\|x_{n-1} + \tau(x_n - x_{n-1}) - x_0\| &= \tau\|x_n - x_{n-1}\| + \|x_{n-1} - x_0\| \\
&\leq \tau\|x_n - x_{n-1}\| + \|x_{n-1} - x_{n-2}\| + \cdots + \|x_1 - x_0\| \\
&\leq \tau(t_n - t_{n-1}) + t_{n-1} - t_0 \\
&= t_{n-1} + \tau(t_n - t_{n-1}) - t_0
\end{aligned}$$

and (1.12), it follows

$$\begin{aligned}
\|x_{n+1} - x_n\| &= \left\|\int_0^1 \mathfrak{S}'\left(x_{n-1} + \tau(x_n - x_{n-1})\right)(x_n - x_{n-1})\,d\tau\right\| \\
&\leq \int_0^1 \mathfrak{g}'\left(t_{n-1} + \tau(t_n - t_{n-1})\right)(t_n - t_{n-1})\,d\tau \\
&= \int_{t_{n-1}}^{t_n} \mathfrak{g}'(\xi)\,d\xi \\
&= \mathfrak{g}(t_n) - \mathfrak{g}(t_{n-1}) \\
&= t_{n+1} - t_n,
\end{aligned}$$

so that

$$\|x_{n+1} - x_0\| \leq \|x_{n+1} - x_n\| + \|x_n - x_{n-1}\| + \cdots + \|x_1 - x_0\|$$
$$\leq t_{n+1} - t_0$$
$$\leq t' - t_0$$
$$= r$$

and $x_{n+1} \in \overline{B(x_0, r)}$.

Moreover, from (1.14), we have

$$\|x_{n+m} - x_n\| \leq \|x_{n+m} - x_{n+m-1}\| + \cdots + \|x_{n+1} - x_n\|$$
$$\leq t_{n+m} - t_{n+m-1} + \cdots + t_{n+1} - t_n$$
$$= t_{n+m} - t_n, \tag{1.15}$$

and, since $\{t_n\}$ is convergent, it is a Cauchy sequence. In addition, $\{x_n\}$ is a Cauchy sequence in the Banach space X and, as a consequence, there exists $x^* = \lim_n x_n$. Taking then into account that $\mathfrak{S}$ is a continuous operator, we obtain

$$x^* = \lim_{n \to +\infty} x_{n+1} = \lim_{n \to +\infty} \mathfrak{S}(x_n) = \mathfrak{S}\left(\lim_{n \to +\infty} x_n\right) = \mathfrak{S}(x^*),$$

so that x^* is a solution of equation (1.9).

Finally, by setting $n = 0$ in (1.15), we have $\|x_m - x_0\| \leq t_m - t_0$, for $m \in \mathbb{N}$, and, by letting $m \to +\infty$, it follows $\|x^* - x_0\| \leq t^* - t_0$. ■

Remark 1.4. Observe in the proof of Theorem 1.3 that we have also proved that the scalar sequence $\{t_n\}$, starting at t_0, is nondecreasing and converges to t^* if the function $\mathfrak{g}$ majorizes the operator $\mathfrak{S}$.

Remark 1.5. By letting $m \to +\infty$ in (1.15), we have $\|x^* - x_n\| \leq t^* - t_n$, for $n \geq 0$, so that we obtain a priori error bounds for Newton's sequence $\{x_n\}$ in the Banach space X from the scalar sequence $\{t_n\}$.

Remark 1.6. Observe that we have not made full use of condition (1.12) in the proof of Theorem 1.3. The proof requires only that $\|\mathfrak{S}'(x)\| \leq \mathfrak{g}'(t)$ for corresponding points of the intervals $[x_{n-1}, x_n]$ and $[t_{n-1}, t_n]$, for $n \in \mathbb{N}$.

Equation (1.9) may have other solutions different from x^* in the closed ball $\overline{B(x_0, r)}$, even if t^* is the unique solution of equation (1.10) in the interval $[t_0, t']$. However, we can establish the following result of uniqueness of solution.

Theorem 1.7. *Under the hypotheses of Theorem 1.3 and $\mathfrak{g}(t') \leq t'$, if equation (1.10) has only the solution t^* in $[t_0, t']$, then equation (1.9) has only one solution x^* in the closed ball $\overline{B(x_0, r)}$. Moreover, the method of successive approximations $x_{n+1} = \mathfrak{S}(x_n)$, $n \geq 0$, starting at any point of $\overline{B(x_0, r)}$, converges to x^*.*

Proof. To begin, we consider $\tilde{t}_{n+1} = \mathfrak{g}(\tilde{t}_n)$, $n \geq 0$, with $\tilde{t}_0 = t'$, and see that the sequence $\{\tilde{t}_n\}$ is nonincreasing and bounded below by t^*, so that it is convergent to t^*.

Using the fact that $\mathfrak{g}$ is nondecreasing, we can prove by mathematical induction that the sequence $\{\tilde{t}_n\}$ is monotone, since the inequalities $\tilde{t}_0 = t' \geq \mathfrak{g}(t') = \mathfrak{g}(\tilde{t}_0) = \tilde{t}_1$ and $\tilde{t}_n \geq \tilde{t}_{n+1}$ imply $\tilde{t}_{n+1} = \mathfrak{g}(\tilde{t}_n) \geq \mathfrak{g}(\tilde{t}_{n+1}) = \tilde{t}_{n+2}$.

As $\tilde{t}_0 = t' \geq t^*$, the inequality $\tilde{t}_1 = \mathfrak{g}(\tilde{t}_0) \geq \mathfrak{g}(t^*) = t^*$ is obvious; and if it has already been proved for $n = j$, then $\tilde{t}_j \geq t^*$, since $\mathfrak{g}$ is nondecreasing and implies $\mathfrak{g}(\tilde{t}_j) \geq \mathfrak{g}(t^*)$; that is, $\tilde{t}_{j+1} \geq t^*$. By mathematical induction, we then have $\tilde{t}_n \geq t^*$, for all $n \geq 0$.

Thus, we have proved that there exists $v = \lim_n \tilde{t}_n$ and, by $\tilde{t}_{n+1} = \mathfrak{g}(\tilde{t}_n)$, $n \geq 0$, and the continuity of $\mathfrak{g}$, v is a solution of equation (1.10); moreover, as t^* is the unique solution of (1.10) in $[t_0, t']$, then $v = t^*$.

Next, we prove that the elements $\tilde{x}_n = \mathfrak{S}(\tilde{x}_{n-1})$, $n \in \mathbb{N}$, are well-defined and form a convergent sequence starting at any point of $\overline{B(x_0, r)}$. First, from $\tilde{x}_0, x_0 \in \overline{B(x_0, r)}$, we obtain respectively $\tilde{x}_{n+1}$ as $\tilde{x}_{n+1} = \mathfrak{S}(\tilde{x}_n)$ and x_{n+1} as $x_{n+1} = \mathfrak{S}(x_n)$, for $n \geq 0$. Besides, from Theorem 1.3, we know that $x_n \in B(x_0, r)$, $n \geq 0$, and $\lim_n x_n = x^*$. Then,

$$\tilde{x}_1 - x_1 = \mathfrak{S}(\tilde{x}_0) - \mathfrak{S}(x_0) = \int_{x_0}^{\tilde{x}_0} \mathfrak{S}'(z)\,dz = \int_0^1 \mathfrak{S}'\left(x_0 + \tau(\tilde{x}_0 - x_0)\right)(\tilde{x}_0 - x_0)\,d\tau \quad (1.16)$$

and, since $\|\tilde{x}_0 - x_0\| \leq r = t' - t_0 = \tilde{t}_0 - t_0$ and

$$\|x_0 + \tau(\tilde{x}_0 - x_0) - x_0\| \leq \tau\|(\tilde{x}_0 - x_0)\| = t_0 + \tau(\tilde{t}_0 - t_0) - t_0,$$

it follows, from (1.16) and condition (1.12), that

$$\|\tilde{x}_1 - x_1\| \leq \int_{t_0}^{\tilde{t}_0} \mathfrak{g}'(\xi)\,d\xi = \mathfrak{g}(\tilde{t}_0) - \mathfrak{g}(t_0) = \tilde{t}_1 - t_1.$$

As a consequence,

$$\|\tilde{x}_1 - x_0\| \leq \|\tilde{x}_1 - x_1\| + \|x_1 - x_0\| \leq \tilde{t}_1 - t_1 + t_1 - t_0 = \tilde{t}_1 - t_0 \leq r$$

and hence $\tilde{x}_1 \in \overline{B(x_0, r)}$.

Suppose that we have already proved that $\tilde{x}_1, \tilde{x}_2, \ldots, \tilde{x}_n \in \overline{B(x_0, r)}$ and

$$\|\tilde{x}_j - x_j\| \leq \tilde{t}_j - t_j, \quad \text{for all} \quad j = 0, 1, \ldots, n. \quad (1.17)$$

Then, as $\tilde{x}_n, x_n \in \overline{B(x_0, r)}$, we obtain

$$\tilde{x}_{n+1} - x_{n+1} = \mathfrak{S}(\tilde{x}_n) - \mathfrak{S}(x_n) = \int_{x_n}^{\tilde{x}_n} \mathfrak{S}'(z)\,dz = \int_0^1 \mathfrak{S}'\left(x_n + \tau(\tilde{x}_n - x_n)\right)(\tilde{x}_n - x_n)\,d\tau$$

and, since $\|\tilde{x}_n - x_n\| \leq \tilde{t}_n - t_n$ and

$$\|x_n + \tau(\tilde{x}_n - x_n) - x_0\| \leq t_n + \tau(\tilde{t}_n - t_n) - t_0,$$

it follows, from condition (1.12), that

$$\|\tilde{x}_{n+1} - x_{n+1}\| \leq \int_{t_n}^{\tilde{t}_n} \mathfrak{g}'(\xi)\,d\xi = \mathfrak{g}(\tilde{t}_n) - \mathfrak{g}(t_n) = \tilde{t}_{n+1} - t_{n+1}.$$

Thus,

$$\|\tilde{x}_{n+1} - x_0\| \leq \|\tilde{x}_{n+1} - x_{n+1}\| + \|x_{n+1} - x_0\| \leq \tilde{t}_{n+1} - t_0 \leq r$$

and so $\tilde{x}_{n+1} \in \overline{B(x_0, r)}$. Then, by mathematical induction on n, we conclude that $x_j \in \overline{B(x_0, r)}$ and (1.17) is true for $j \geq 0$.

On the other hand, we have that the sequence $\{t_n\}$ is nondecreasing and convergent to t^*, the sequence $\{\tilde{t}_n\}$ is nonincreasing and convergent to t^* and the sequence $\{x_n\}$ is convergent to x^*. Then, from $\|\tilde{x}_n - x_n\| \leq \tilde{t}_n - t_n$, we obtain $\lim_n \tilde{x}_n = \lim_n x_n = x^*$, so that the sequence $\tilde{x}_{n+1} = \mathfrak{S}(\tilde{x}_n)$, $n \geq 0$, converges to x^* whatever the initial approximation $\tilde{x}_0 \in \overline{B(x_0, r)}$ is used.

The uniqueness of solution of equation (1.9) is immediate from the last fact, since if y^* is other solution of equation (1.9) different from x^* and choose $\tilde{x}_0 = y^*$, we obviously have $\tilde{x}_n = y^*$, for $n \geq 0$, so that $\{\tilde{x}_n\}$ is constant. Then, as $\lim_n \tilde{x}_n = x^*$, it follows $x^* = y^*$ and the proof is complete. ∎

Remark 1.8. In Theorem 1.7, it has been proved that $\overline{B(x_0, r)}$ is a ball of convergence for the sequence defined from the operator $\mathfrak{S}$, since the corresponding method of successive approximations is convergent starting at any point of $\overline{B(x_0, r)}$.

In addition, Kantorovich first considers an operator F defined on the ball $B(x_0, R)$ and a scalar function f on the interval $[t_0, t']$, supposes that there exists the operator $\Gamma_0 = [F'(x_0)]^{-1}$ with $\|\Gamma_0\| \leq -\frac{1}{f'(t_0)}$, takes into account $\mathfrak{S}(x) = x - \Gamma_0 F(x)$ and $\mathfrak{g}(t) = t - \frac{f(t)}{f'(t_0)}$, and proves that the sequence $x_{n+1} = x_n - \Gamma_0 F(x_n)$, $n \geq 0$, generated by the modified Newton method, is convergent from the convergence of the scalar sequence $t_{n+1} = t_n - \frac{f(t_n)}{f'(t_0)}$, $n \geq 0$, if the function $\mathfrak{g}$ majorizes the operator $\mathfrak{S}$. Next, through an inductive process, Kantorovich extends this study to the convergence of Newton's method.

After that, we start studying the semilocal convergence of Newton's method to approximate a solution of the operator equation $F(x) = 0$, where $F : B(x_0, R) \subseteq X \longrightarrow Y$ is a twice continuously Fréchet differentiable operator defined on the ball $B(x_0, R)$. For this, we consider $f : [t_0, t'] \longrightarrow \mathbb{R}$, with $t' = t_0 + r$ and such that f is twice differentiable on $[t_0, t']$, and a closed ball $\overline{B(x_0, r)}$ with $r < R$, choose

$$\mathfrak{S}(x) = x - [F'(x)]^{-1}F(x) \quad \text{and} \quad \mathfrak{g}(t) = t - \frac{f(t)}{f'(t)},$$

and see the modified Newton methods as the methods of successive approximations applied to the operators

$$\hat{\mathfrak{S}}(x) = x - [F'(x_0)]^{-1}F(x) \quad \text{and} \quad \hat{\mathfrak{g}}(t) = t - \frac{f(t)}{f'(t_0)}.$$

First, we obtain a semilocal convergence result for the modified Newton method in the Banach space X. Note that condition (1.12) is easier to prove for the operator $\hat{\mathfrak{S}}$ and the function $\hat{\mathfrak{g}}$ than for $\mathfrak{S}$ and $\mathfrak{g}$. Second, we extend the result obtained for the modified Newton method to Newton's method.

Theorem 1.9. *Let $F : B(x_0, R) \subseteq X \longrightarrow Y$ be a twice continuously Fréchet differentiable operator defined on the ball $B(x_0, R)$ of a Banach space X with values in a Banach space Y. Suppose that there exists $f \in C^2([t_0, t'])$ with $t' - t_0 = r < R$ and such that the following conditions are satisfied:*

(a) There exists $\Gamma_0 = [F'(x_0)]^{-1} \in \mathcal{L}(Y, X)$, $\|\Gamma_0\| \leq -\dfrac{1}{f'(t_0)}$ and $\|\Gamma_0 F(x_0)\| \leq -\dfrac{f(t_0)}{f'(t_0)}$.

(b) $\|F''(x)\| \leq f''(t)$ for $\|x - x_0\| \leq t - t_0$, $x \in B(x_0, R)$ and $t \in [t_0, t']$.

If the equation $f(t) = 0$ has only one solution t^* in $[t_0, t']$, then the methods of successive approximations given by $\widehat{x}_{n+1} = \widehat{\mathfrak{S}}(\widehat{x}_n)$, $n \geq 0$, and $\widehat{t}_{n+1} = \widehat{\mathfrak{g}}(\widehat{t}_n)$, $n \geq 0$, starting respectively at $\widehat{x}_0 = x_0$ and $\widehat{t}_0 = t_0$, converge respectively to x^* and t^*, where x^* is a solution of the equation $F(x) = 0$, and such that $\|x^* - x_0\| \leq t^* - t_0$.

Proof. We start proving that the function $\widehat{\mathfrak{g}}$ majorizes the operator $\widehat{\mathfrak{S}}$. On the one hand, from condition (a), it follows $\|\widehat{\mathfrak{S}}(x_0) - x_0\| \leq \widehat{\mathfrak{g}}(t_0) - t_0$. On the other hand, from

$$\widehat{\mathfrak{S}}'(x) = I - \Gamma_0 F'(x) \qquad \text{and} \qquad \widehat{\mathfrak{S}}''(x) = -\Gamma_0 F''(x),$$

it follows

$$\widehat{\mathfrak{S}}'(x) = \widehat{\mathfrak{S}}'(x) - \widehat{\mathfrak{S}}'(x_0) = \int_{x_0}^x \widehat{\mathfrak{S}}''(z)\, dz = -\int_{x_0}^x \Gamma_0 F''(z)\, dz,$$

and

$$\|\widehat{\mathfrak{S}}'(x)\| = \left\| -\int_{x_0}^x \Gamma_0 F''(z)\, dz \right\| = \left\| -\int_0^1 \Gamma_0 F''(x_0 + \tau(x - x_0))\, d\tau (x - x_0) \right\|.$$

Then, as

$$\|x_0 + \tau(x - x_0) - x_0\| \leq \tau \|x - x_0\| \leq \tau(t - t_0) = t_0 + \tau(t - t_0) - t_0,$$

from condition (b), it follows

$$\|F''(x_0 + \tau(x - x_0))\| \leq f''(t_0 + \tau(t - t_0))$$

and, taking again into account condition (a), we have

$$\begin{aligned}
\|\widehat{\mathfrak{S}}'(x)\| &\leq \int_0^1 \left(-\frac{1}{f'(t_0)} \right) f''(t_0 + \tau(t - t_0))\, d\tau(t - t_0) \\
&= -\frac{1}{f'(t_0)} \int_{t_0}^t f''(\xi)\, d\xi \\
&= -\frac{1}{f'(t_0)} (f'(t) - f'(t_0)) \\
&= 1 - \frac{f'(t)}{f'(t_0)} \\
&= \widehat{\mathfrak{g}}'(t).
\end{aligned}$$

Therefore, the function $\widehat{\mathfrak{g}}$ majorizes the operator $\widehat{\mathfrak{S}}$ and the proof is completed by using Theorem 1.3. ∎

Remark 1.10. Observe that, as in Theorem 1.3, we have

$$\|x^* - \widehat{x}_n\| \leq t^* - \widehat{t}_n, \quad n \geq 0, \tag{1.18}$$

taking into account $\widehat{x}_0 = x_0$ and $\widehat{t}_0 = t_0$ as starting points for $\widehat{x}_{n+1} = \widehat{\mathfrak{S}}(\widehat{x}_n)$, $n \geq 0$, and $\widehat{t}_{n+1} = \widehat{\mathfrak{g}}(\widehat{t}_n)$, $n \geq 0$, respectively.

Now, from Theorem 1.7, we establish the following result of uniqueness of solution for the modified Newton method.

Theorem 1.11. *Under the hypotheses of Theorem 1.9 and condition $f(t') \leq 0$, if the equation $f(t) = 0$ has only one solution in $[t_0, t']$, then the equation $F(x) = 0$ has only one solution in the closed ball $\overline{B(x_0, r)}$.*

Proof. If $f(t') \leq 0$, we have

$$\widehat{g}(t') = t' - \frac{f(t')}{f'(t_0)} \leq t',$$

so that the additional condition of Theorem 1.7 is satisfied. Then, from Theorem 1.7, it follows the uniqueness of solution. ∎

Remark 1.12. Since t^* is the smallest positive solution of the equation $f(t) = 0$, we can apply Theorem 1.11 when $t' = t^*$. Therefore, the uniqueness of solution x^* is always guaranteed in the closed ball $\overline{B(x_0, t^* - t_0)}$, which is also the domain of existence of solution.

The semilocal convergence of Newton's method can be then established from the semilocal convergence of the modified Newton method established in Theorem 1.9.

Theorem 1.13. *Under the hypotheses of Theorem 1.9, Newton's method, starting at $x_0 \in B(x_0, R)$, converges to a solution x^* of the equation $F(x) = 0$.*

Proof. We consider

$$t_{n+1} = N_f(t_n) = t_n - \frac{f(t_n)}{f'(t_n)}, \quad n \geq 0, \quad \text{with } t_0 \text{ given,}$$

and Newton's sequence $\{x_n\}$ given in (1.1).

As the first step of Newton's method and the modified Newton method is the same, then x_1 is well defined and $x_1 \in \overline{B(x_0, r)}$.

We now verify that the conditions of Theorem 1.9 are not violated when x_0 is replaced by x_1 and t_0 by t_1.

Taking into account

$$I - \Gamma_0 F'(x_1) = -\Gamma_0 (F'(x_1) - F'(x_0))$$
$$= -\int_{x_0}^{x_1} \Gamma_0 F''(z) \, dz$$
$$= -\int_0^1 \Gamma_0 F'' (x_0 + \tau(x_1 - x_0)) \, d\tau(x_1 - x_0),$$

and

$$\|x_0 + \tau(x_1 - x_0) - x_0\| \leq \tau\|x_1 - x_0\| \leq \tau(t_1 - t_0) = t_0 + \tau(t_1 - t_0) - t_0,$$

it follows, from condition (b) of Theorem 1.9, that

$$\|I - \Gamma_0 F'(x_1)\| \leq \|\Gamma_0\| \int_0^1 \|F'' (x_0 + \tau(x_1 - x_0))\| \, dt \|x_1 - x_0\|$$
$$\leq -\frac{1}{f'(t_0)} \int_0^1 f'' (t_0 + \tau(t_1 - t_0)) \, d\tau(t_1 - t_0)$$
$$= -\frac{1}{f'(t_0)} \int_{t_0}^{t_1} f''(\xi) \, d\xi$$
$$= -\frac{1}{f'(t_0)} (f'(t_1) - f'(t_0))$$
$$= 1 - \frac{f'(t_1)}{f'(t_0)}.$$

From conditions (a) and (b) of Theorem 1.9, we have $f''(t) \geq 0$ in $[t_0, t']$ and $f(t_0) \geq 0$. Besides, the function f cannot have a minimum to the left of the solution t^*, since t^* is the lowest positive zero of f, so that $f'(t) < 0$ in $[t_0, t^*]$. Now, by the Mean-Value Theorem, there exists $\theta \in (t_0, t^*)$ such that

$$t_1 - t^* = N_f(t_0) - N_f(t^*) = N_f'(\theta)(t_0 - t^*) = \frac{f(\theta) f''(\theta)}{f'(\theta)^2}(t_0 - t^*)$$

and then $t_1 \leq t^*$. Therefore, $t_1 \in [t_0, t^*]$ and $1 - \frac{f'(t_1)}{f'(t_0)} < 1$.

By the Banach lemma on invertible operators, the operator $[F'(x_1)]^{-1}$ exists and

$$\|\Gamma_1\| = \|[F'(x_1)]^{-1}\| \leq \frac{\|\Gamma_0\|}{1 - \left(1 - \frac{f'(t_1)}{f'(t_0)}\right)} \leq \frac{-\frac{1}{f'(t_0)}}{\frac{f'(t_1)}{f'(t_0)}} = -\frac{1}{f'(t_1)}.$$

Moreover, as

$$F(x_1) = F(x_0) + F'(x_0)(x_1 - x_0) + \int_{x_0}^{x_1} F''(z)(x_1 - z)\, dz$$

$$= \int_0^1 F''(x_0 + \tau(x_1 - x_0))(x_1 - x_0)^2(1 - \tau)\, dz,$$

and $\|x_1 - x_0\| \leq -\frac{f(t_0)}{f'(t_0)} = t_1 - t_0$, we have

$$\|x_0 + \tau(x_1 - x_0) - x_0\| \leq \tau\|x_1 - x_0\| \leq \tau(t_1 - t_0) = t_0 + \tau(t_1 - t_0) - t_0.$$

Then, from condition (b) of Theorem 1.9, it follows

$$\|F'(x_1)\| \leq \int_0^1 f''(t_0 + \tau(t_1 - t_0))(1 - \tau)(t_1 - t_0)^2\, d\tau$$

$$= \int_{t_0}^{t_1} f''(\xi)(t_1 - \xi)\, d\xi$$

$$= f(t_0) + f'(t_0)(t_1 - t_0) + \int_{t_0}^{t_1} f''(\xi)(t_1 - \xi)\, d\xi$$

$$= f(t_1)$$

and, as a consequence,

$$\|\Gamma_1 F(x_1)\| \leq \|\Gamma_1\|\|F(x_1)\| \leq -\frac{f(t_1)}{f'(t_1)}.$$

Moreover, if $\|x - x_1\| \leq t - t_1$, then

$$\|x - x_0\| \leq \|x - x_1\| + \|x_1 - x_0\| \leq t - t_1 + t_1 - t_0 = t - t_0,$$

and

$$\|F''(x)\| \leq f''(t) \quad \text{for} \quad \|x - x_1\| \leq t - t_1.$$

Furthermore, since $t_1 \leq t^*$, it is clear that $f(t) = 0$ has a solution in $[t_1, t']$ and, from

$$t_1 - t_0 = -\frac{f(t_0)}{f'(t_0)} \geq 0,$$

it follows $t_0 \leq t_1 \leq t^*$.

Therefore, the hypotheses of Theorem 1.9 are also satisfied when we substitute x_1 and t_1 for x_0 and t_0, respectively. This allows us to continue the successive determination of the elements x_n and t_n, so that $x_n \in \overline{B(x_0, r)}$ and

$$\|x_{n+1} - x_n\| = \|\Gamma_n F(x_n)\| \leq \|\Gamma_n\| \|F(x_n)\| \leq -\frac{f(t_n)}{f'(t_n)} = t_{n+1} - t_n, \qquad (1.19)$$

for all $n \geq 0$.

On the other hand, as $t_n \leq t_{n+1}$ and $t_n \leq t^*$, for all $n \geq 0$, there exists $v = \lim_n t_n \leq t^*$. Besides, as $v = \lim_n t_n = \lim_n N_f(t_{n-1}) = N_f(v)$, then $f(v) = 0$ and, therefore, $v = t^*$. In addition, from (1.19), it also follows that there exists $x^* = \lim_n x_n$.

Now, from the algorithm of Newton's method (1.1), we have

$$F(x_n) + F'(x_n)(x_{n+1} - x_n) = 0, \qquad n \geq 0, \qquad (1.20)$$

and, by the Mean-Value Theorem, we obtain, for $x \in \overline{B(x_0, r)}$, that

$$\|F'(x) - F'(x_0)\| \leq \sup_{0 < \theta < 1} \|F''(x_0 + \theta(x - x_0))\| \|x - x_0\| \leq \left(\max_{[t_0, t']} \{f''(t)\} \right) \cdot r,$$

since

$$\|x_0 + \theta(x - x_0) - x_0\| = \theta \|x - x_0\| \leq \theta r = \theta(t' - t_0) = t_0 + \theta(t' - t_0) - t_0,$$

so that $\|F'(x_n)\|$ is bounded:

$$\|F'(x_n)\| \leq \|F'(x_0)\| + \left(\max_{[t_0, t']} \{f''(t)\} \right) \cdot r.$$

Then, taking limits in (1.20) and bearing in mind that F is continuous, we obtain $F(x^*) = 0$, so that x^* is a solution of $F(x) = 0$. ∎

Remark 1.14. Note that we also have the following estimates $\|x^* - x_n\| \leq t^* - t_n$, $n \geq 0$. This is a consequence of (1.18), since x_n and t_n can be seen as first approximations in the modified Newton method with initial approximations x_{n-1} and t_{n-1}, respectively.

Next, from Theorem 1.9, we establish the following result of uniqueness of solution for Newton's method.

Theorem 1.15. *Under the hypotheses of Theorem 1.9, the equation $F(x) = 0$ has only one solution in the closed ball $\overline{B(x_0, t^* - t_0)}$, where t^* is the unique solution of $f(t) = 0$ in $[t_0, t']$.*

Proof. If $r = t^* - t_0$, then $t' = t^*$ and $f(t') = f(t^*) = 0$. Then, from Theorem 1.11, it follows the uniqueness of solution. ∎

Note that the semilocal convergence results given by Theorems 1.9 and 1.13 are often difficult to make direct use, since they contain the undefined function f. To know what the function f is, some conditions on the operator F are required. For this, the following result acquires importance.

Theorem 1.16. *Let $F : B(x_0, R) \subseteq X \longrightarrow Y$ be a twice continuously Fréchet differentiable operator defined on an open ball $B(x_0, R)$ of a Banach space X with values in a Banach space Y. If conditions (W1)-(W2)-(W3)-(W4) are satisfied, then Newton's method (and the modified Newton method), starting at x_0, converges to a solution x^* of the equation $F(x) = 0$.*

Proof. To apply Theorem 1.13, we need to know the scalar function $f : [t_0, t'] \longrightarrow \mathbb{R}$. As quadratic polynomials are the most basic functions that satisfy condition (W3), we can find f by solving a problem of interpolation fitting from fixing the three coefficients of the quadratic polynomial by conditions (W1)-(W2)-(W3). So, we do

$$f''(t) = M, \qquad -\frac{1}{f'(t_0)} = \beta, \qquad -\frac{f(t_0)}{f'(t_0)} = \eta$$

and obtain

$$f(t) = \frac{M}{2}(t - t_0)^2 - \frac{t - t_0}{\beta} + \frac{\eta}{\beta}. \tag{1.21}$$

As the two zeros of the last polynomial are $t_0 + \frac{1 \pm \sqrt{1-2h}}{h}\eta$, if $r \geq t_0 + \frac{1-\sqrt{1-2h}}{h}\eta = t^*$, then $t' = r + t_0 \geq s^* + t_0 = t^*$, where $s^* = \frac{1-\sqrt{1-2h}}{h}\eta$. Therefore, (1.21) satisfies all the conditions of Theorem 1.13 and the thesis of the theorem is proved. ∎

In addition, we can also establish the following result on the uniqueness of solution.

Theorem 1.17. *Under the hypotheses of Theorem 1.16, the equation $F(x) = 0$ has only one solution in the closed ball $\overline{B(x_0, t^* - t_0)}$, where $t^* = t_0 + \frac{1-\sqrt{1-2h}}{h}\eta$.*

Proof. As t^* is the unique zero of quadratic polynomial (1.21) in $[t_0, t^*]$, then the thesis of the theorem follows from Theorem 1.15. ∎

As we have just seen, Kantorovich also gives a result of uniqueness of solution, which is obtained from Theorem 1.11 for the modified Newton method with the condition $f(t') = 0$, that is: $t' = t^*$.

Remark 1.18. (Property of translation of Kantorovich's polynomial.) Notice that we have considered any $t_0 \geq 0$ in the above-mentioned, but Kantorovich considers $t_0 = 0$, so that the quadratic polynomial f given in (1.21) is reduced to $p(t) = \frac{M}{2}t^2 - \frac{t}{\beta} + \frac{\eta}{\beta}$, which is usually known as *Kantorovich's polynomial*. This is a consequence of the fact that $p(t) = f(t + t_0)$. In addition, the sequence $\{t_{n+1} = N_f(t_n)\}_{n \geq 0}$, for any $t_0 > 0$, satisfies $t_{n+1} = N_f(t_n) = t_0 + N_p(s_n)$, $n \geq 0$, where $s_{n+1} = N_p(s_n)$ with $s_0 = 0$, since we have, for $t_0 \geq 0$ and $s_0 = 0$,

$$t_0 + s_{n+1} = t_0 + N_p(s_n) = t_0 + s_n - \frac{p(s_n)}{p'(s_n)} = t_0 + s_n - \frac{f(s_n + t_0)}{f'(s_n + t_0)} = t_n - \frac{f(t_n)}{f'(t_n)} = N_f(t_n) = t_{n+1},$$

for all $n \geq 0$. Therefore, the real sequences $\{t_n\}$ and $\{s_n\}$ given by Newton's method when they are constructed from f and p, respectively, can be obtained, one from the other, by translation. Besides, $t_{n+1} - t_n = s_{n+1} - s_n$, for all $n \geq 0$, and all the results obtained previously are independent of the value $t_0 \geq 0$, so that we choose $t_0 = 0$ because, in practice, it is the most favorable situation.

On the other hand, if we consider Newton's method given by

$$t_{n+1} = \mathfrak{g}(t_n) = t_n - \frac{p(t_n)}{p'(t_n)}, \quad n \geq 0,$$

starting at $t_0 = t'$, with $t' > t^*$, we do not obtain a nonincreasing sequence convergent to t^*, which is the reasoning used in Theorem 1.7 for the uniqueness of solution. So, although $p(t') < 0$, the uniqueness of solution is not deduced directly from Theorem 1.11, unless the modified Newton method is considered, as we have seen. However, using other arguments, we see later that the uniqueness of solution can be extended until $B(x_0, t^{**} - t_0) \cap B(x_0, R)$, where $t^{**} = t_0 + \frac{1+\sqrt{1-2h}}{h}\eta$.

1.1.3 The "method of majorizing sequences"

Next, following Ortega [63], who gave a proof of the Newton-Kantorovich theorem, which is a modification of the previous approach of Kantorovich, we give a proof of the theorem which is also a modification of the Kantorovich approach and is, we believe, easier to understand and present.

Notice that Kantorovich considers the open ball $B(x_0, R)$ as domain of the operator F, but this is very restrictive. Then, we consider an open convex region Ω of the Banach space X as domain of the operator F, so that we can follow the same arguments of Kantorovich.

Majorizing sequence

From the study of the semilocal convergence of Newton's method performed by Kantorovich through the majorant principle, turns clearly out the difficulty of its application to other iterative methods as a result of the concept of majorant function of an operator. However, there is one fact that is distinguised from Kantorovich's theory: estimates

$$\|x^* - x_n\| \leq t^* - t_n, \quad n \geq 0,$$

given in Remark 1.5. From this fact and the concept of *majorizing sequence*, which is defined below, we can prove the semilocal convergence of Newton's method more easily. This concept also serves to introduce a modification of the majorant principle of Kantorovich, whose application is easier and allows modifying the classic conditions of Kantorovich. So, we introduce the concept of majorizing sequence and show how it is used to prove the convergence of sequences in Banach spaces.

Definition 1.19. *If $\{x_n\}$ is a sequence in a Banach space X and $\{t_n\}$ is a real sequence, then $\{t_n\}$ is a majorizing sequence of $\{x_n\}$ if $\|x_{n+1} - x_n\| \leq t_{n+1} - t_n$, for all $n \geq 0$.*

From the last inequality, we observe that the sequence $\{t_n\}$ is nondecreasing. The interest of the majorizing sequence is that the convergence of the sequence $\{x_n\}$ in the Banach space X is deduced from the convergence of the scalar sequence $\{t_n\}$, as we see in the following result [63].

Theorem 1.20. *Let $\{x_n\}$ be a sequence in a Banach space X and $\{t_n\}$ a majorizing sequence of $\{x_n\}$. Then, if $\{t_n\}$ converges to $t^* \in \mathbb{R}$, there exists a $x^* \in X$ such that $x^* = \lim_n x_n$ and $\|x^* - x_n\| \leq t^* - t_n$, for $n = 0, 1, 2, \ldots$*

The proof of Theorem 1.20 is immediate from

$$\|x_{n+m} - x_n\| \leq \sum_{i=1}^{m} \|x_{n+i} - x_{n+i-1}\| \leq t_{n+m} - t_n \leq t^* - t_n,$$

which proves that $\{x_n\}$ is a Cauchy sequence in the Banach space X.

From the definition of majorizing sequence and Theorem 1.20, Kantorovich proves the Newton-Kantorovich theorem. For this, a majorizing sequence is constructed from conditions (W1)-(W2)-(W3) of the Newton-Kantorovich theorem, by applying Newton's method,

$$s_0 = 0, \qquad s_{n+1} = N_p(s_n) = s_n - \frac{p(s_n)}{p'(s_n)}, \qquad n = 0, 1, 2, \ldots, \tag{1.22}$$

to Kantorovich's polynomial

$$p(s) = \frac{M}{2}s^2 - \frac{s}{\beta} + \frac{\eta}{\beta}. \tag{1.23}$$

Note that (1.23) has two positive zeros $s^* = \frac{1-\sqrt{1-2h}}{h}\eta$ and $s^{**} = \frac{1+\sqrt{1-2h}}{h}\eta$ such that $s^* \leq s^{**}$ if $h \leq \frac{1}{2}$. Moreover, we consider $p(s)$ in some interval $[0, s']$ taking into account that $s^* \leq s^{**} < s'$.

Theorem 1.21. *The real sequence $\{s_n\}$ given in (1.22) is nondecreasing and converges to the unique zero s^* of polynomial (1.23) in (s_0, α), where α is the minimum of polynomial (1.23).*

Proof. We prove first that the sequence $\{s_n\}$ is nondecreasing sequence and convergent to s^*. To this end, we note that the sequence $\{s_n\}$ is monotone and bounded.

As $p(t) > 0$, $p'(t) < 0$ and $p''(t) > 0$ in (s_0, α), it is easy to prove by mathematical induction that $s_n \leq s^*$, for all $n \geq 0$. As $p(s_0) = p(0) = \frac{\eta}{\beta} > 0$, then $s_0 \leq s^*$ and

$$s_1 - s^* = N_p(s_0) - N_p(s^*) = N_p'(\theta_0)(s_0 - s^*), \qquad \theta_0 \in (s_0, s^*).$$

Now, from $N_p'(t) = \frac{p(t)p''(t)}{p'(t)^2} > 0$ in $[s_0, s^*)$, since $p(t) > 0$ and $p''(t) > 0$ in $[s_0, s^*)$, we have $s_1 < s^*$. If we now assume $s_n < s^*$, it follows in the same way that

$$s_{n+1} - s^* = N_p(s_n) - N_p(s^*) = N_p'(\theta_n)(s_n - s^*), \qquad \theta_n \in (s_n, s^*),$$

and $s_{n+1} < s^*$, so that $s_n < s^*$ is true for all positive integer n by mathematical induction.

On the other hand, since $p(t) > 0$ and $p'(t) < 0$ in $[s_0, s^*)$, we have

$$s_n - s_{n-1} = -\frac{p(s_{n-1})}{p'(s_{n-1})} \geq 0$$

and $\{s_n\}$ is a nondecreasing sequence.

Therefore, we infer that there exists $v = \lim_n s_n \in [s_0, s^*]$ and, by the continuity of N_p, v is a solution of $p(s) = 0$; moreover, as s^* is the unique zero of $p(s)$ in $[s_0, s^*]$, then $v = s^*$. ∎

Remark 1.22. On the other hand, observe that the radii of existence ρ^* and uniqueness ρ^{**} of the solution x^* of $F(x) = 0$, given in the Newton-Kantorovich theorem, fit in with s^* and s^{**}, respectively.

Now, we prove the Newton-Kantorovich theorem by using a modification of the majorant principle of Kantorovich, which is based on the concept of majorizing sequence. For this, we consider polynomial (1.23) and suppose the following conditions:

(H1) there exists $\Gamma_0 = [F'(x_0)]^{-1} \in \mathcal{L}(Y, X)$, for some $x_0 \in \Omega$, with $\|\Gamma_0\| \leq \beta$,

(H2) $\|\Gamma_0 F(x_0)\| \leq \eta$,

(H3) $\|F''(x)\| \leq M, \ x \in \Omega$,

(H4) $h = M\beta\eta \leq \frac{1}{2}$,

where Ω is a non-empty open convex domain of the Banach space X. Conditions (H1)-(H2)-(H3)-(H4) are commonly known as *the classic conditions of Kantorovich* and are conditions (W1)-(W2)-(W3)-(W4) extended to the domain Ω.

Theorem 1.23. *Let $F : \Omega \subseteq X \longrightarrow Y$ be a twice continuously Fréchet differentiable operator defined on a nonempty open convex domain Ω of a Banach space X with values in a Banach space Y. Suppose that conditions (H1)-(H2)-(H3)-(H4) are satisfied and $B(x_0, s^*) \subset \Omega$ with $s^* = \frac{1-\sqrt{1-2h}}{h}\,\eta$. Then, Newton's sequence defined in (1.1) and starting at x_0 converges to a solution x^* of the equation $F(x) = 0$ and the solution x^* and the iterates x_n belong to $\overline{B(x_0, s^*)}$, for all $n = 0, 1, 2, \ldots$ Moreover, if $h = M\beta\eta < \frac{1}{2}$, the solution x^* is unique in $B(x_0, s^{**}) \cap \Omega$, where $s^{**} = \frac{1+\sqrt{1-2h}}{h}\,\eta$, and, if $h = \frac{1}{2}$, x^* is unique in $\overline{B(x_0, s^*)}$. Furthermore, we have the error estimates given in (1.2).*

Proof. First, x_1 is well-defined, since the operator $\Gamma_0 = [F'(x_0)]^{-1}$ exists by the hypotheses. Moreover,

$$\|x_1 - x_0\| = \|\Gamma_0 F(x_0)\| \leq \|\Gamma_0\| \|F(x_0)\| \leq -\frac{p(s_0)}{p'(s_0)} = s_1 - s_0 < s^*,$$

where p is defined in (1.23), and $x_1 \in B(x_0, s^*)$.

Next, taking into account that

$$\|I - \Gamma_0 F'(x_1)\| \leq \|\Gamma_0\| \|F'(x_0) - F'(x_1)\|$$

$$\leq M\beta_0 \|x_1 - x_0\|$$

$$\leq -\frac{M}{p'(s_0)}(s_1 - s_0)$$

$$= -\frac{1}{p'(s_0)}(p'(s_1) - p'(s_0))$$

$$= 1 - \frac{p'(s_1)}{p'(s_0)}$$

$$< 1,$$

since $p'(s_1) > 0$ and $p'(s_1) > p'(s_0)$, it follows, by the Banach lemma on invertible operators, that the operator $\Gamma_1 = [F'(x_1)]^{-1}$ exists and

$$\|\Gamma_1\| \leq -\frac{1}{p'(s_1)}.$$

Hence, x_2 is well-defined.

Besides, since

$$F(x_1) = F(x_0) + F'(x_0)(x_1 - x_0) + \int_{x_0}^{x_1} F''(z)(x_1 - z)\,dz$$

$$= \int_{x_0}^{x_1} F''(z)(x_1 - z)\,dz$$

$$= \int_0^1 F''(x_0 + \tau(x_1 - x_0))(x_1 - x_0)^2(1 - \tau)\,d\tau,$$

we have

$$\|F(x_1)\| \leq \int_0^1 \|F''(x_0 + \tau(x_1 - x_0))\|\,(1 - \tau)\,d\tau\|x_1 - x_0\|^2$$

$$\leq \frac{M}{2}\|x_1 - x_0\|^2$$

$$\leq \frac{M}{2}(s_1 - s_0)^2$$

$$= \int_{s_0}^{s_1} p''(\tau)(s_1 - \tau)\,d\tau$$

$$= p(s_1),$$

and it follows that

$$\|x_2 - x_1\| = \|\Gamma_1 F(x_1)\| \leq \|\Gamma_1\|\|F(x_1)\| \leq -\frac{p(s_1)}{p'(s_1)} = s_2 - s_1,$$

$$\|x_2 - x_0\| \leq \|x_2 - x_1\| + \|x_1 - x_0\| \leq (s_2 - s_1) + (s_1 - s_0) = s_2 - s_0 < s^*.$$

Thus, $x_2 \in B(x_0, s^*)$.

If we assume

$$\|\Gamma_n\| \leq -\frac{1}{p'(s_n)}, \tag{1.24}$$

$$\|x_{n+1} - x_n\| \leq -\frac{p(s_n)}{p'(s_n)} = s_{n+1} - s_n, \tag{1.25}$$

$$\|x_{n+1} - x_0\| \leq s_{n+1} - s_0 < s^*, \tag{1.26}$$

where the operator $\Gamma_n = [F'(x_n)]^{-1}$ exists, it follows in the same way that the operator $\Gamma_{n+1} = [F'(x_{n+1})]^{-1}$ exists and

$$\|\Gamma_{n+1}\| \leq -\frac{1}{p'(s_{n+1})},$$

$$\|x_{n+2} - x_{n+1}\| \leq -\frac{p(s_{n+1})}{p'(s_{n+1})} = s_{n+2} - s_{n+1},$$

$$\|x_{n+2} - x_0\| \leq \|x_{n+2} - x_{n+1}\| + \|x_{n+1} - x_0\| \leq s_{n+2} - s_0 < s^*,$$

so that (1.24), (1.25) and (1.26) are true for all positive integers n by mathematical induction. As a consequence, the sequence $\{x_n\}$ is well-defined and $x_n \in B(x_0, s^*)$, for all $n \geq 0$.

Since $\lim_n s_n = s^*$, then $\{s_n\}$ is a Cauchy sequence, so that $\{x_n\}$ is also a Cauchy sequence and then convergent. So, $\lim_n x_n = x^*$ and

$$\|x^* - x_n\| \leq s^* - s_n, \quad n = 0, 1, 2, \ldots \tag{1.27}$$

Moreover, from (1.1), we have

$$F(x_n) + F'(x_n)(x_{n+1} - x_n) = 0, \quad n = 0, 1, 2, \ldots, \tag{1.28}$$

and taking into account that the sequence $\{\|F'(x_n)\|\}$ is bounded, since

$$\begin{aligned}
\|F'(x_n)\| &\leq \|F'(x_0)\| + \|F'(x_n) - F'(x_0)\| \\
&\leq \|F'(x_0)\| + M\|x_n - x_0\| \\
&\leq \|F'(x_0)\| + M(s_n - s_0) \\
&\leq \|F'(x_0)\| + M(s^* - s_0),
\end{aligned}$$

we obtain $F(x^*) = 0$ by the continuity of F. As a consequence, x^* is a solution of $F(x) = 0$.

Next, we prove the uniqueness of the solution x^*. For this, we suppose that there exists another solution $y^* \in B(x_0, s^{**}) \cap \Omega$ of $F(x) = 0$ such that $y^* \neq x^*$ and follow a different technique from that used before in the proof of Theorem 1.1. Consider

$$F(y^*) - F(x^*) = \int_{x^*}^{y^*} F'(z)\, dz = \int_0^1 F'(x^* + \tau(y^* - x^*))(y^* - x^*)\, d\tau = 0 \tag{1.29}$$

and the operator $J = \int_0^1 F'(x^* + \tau(y^* - x^*))\, d\tau$. Since

$$\begin{aligned}
\|I - \Gamma_0 J\| &\leq \|\Gamma_0\|\|F'(x_0) - J\| \\
&\leq \|\Gamma_0\| \int_0^1 \|F'(x_0) - F'(x^* + \tau(y^* - x^*))\|\, d\tau \\
&\leq M\beta \int_0^1 ((1 - \tau)\|x^* - x_0\| + \tau(\|y^* - x_0\|))\, d\tau \\
&< \frac{1}{2} M\beta(s^* + s^{**}) \\
&= 1,
\end{aligned}$$

the operator J is invertible, by the Banach lemma on invertible operators, and then $x^* = y^*$.

Finally, we establish inequality (1.2) giving the rate of convergence of Newton's method (see [55]). First, we consider method (1.22), write

$$\eta_n = -\frac{p(s_n)}{p'(s_n)}, \quad M_n = -\frac{p''(s_n)}{p'(s_n)} = -\frac{2M}{p'(s_n)}, \quad h_n = M_n \eta_n, \quad n \geq 0,$$

and express η_n, M_n and h_n in terms of the corresponding terms with suffix $n - 1$. For this, we note that

$$s_{n+1} - s_n = -\frac{p(s_n)}{p'(s_n)} = \eta_n, \quad n \geq 0, \tag{1.30}$$

and, by Taylor's series for a polynomial of degree two, we obtain

$$\eta_n = -\frac{p(s_n)}{p'(s_n)}$$

$$= -\frac{1}{p'(s_n)}p(s_{n-1} + \eta_{n-1})$$

$$= -\frac{1}{p'(s_n)}\left(p(s_{n-1}) + p'(s_{n-1})\eta_{n-1} + \frac{1}{2}p''(s_{n-1})\eta_{n-1}^2\right)$$

$$= -\frac{1}{p'(s_n)}\left(-p'(s_{n-1})\eta_{n-1} + p'(s_{n-1})\eta_{n-1} + M\eta_{n-1}^2\right)$$

$$= -\frac{1}{p'(s_n)}M\eta_{n-1}^2$$

$$= -\frac{1}{2}\frac{p'(s_{n-1})}{p'(s_n)}\frac{2M}{p'(s_{n-1})}\eta_{n-1}^2$$

$$= \frac{1}{2}\frac{p'(s_{n-1})}{p'(s_n)}M_{n-1}\eta_{n-1}^2.$$

However,

$$\frac{p'(s_n)}{p'(s_{n-1})} = \frac{p'(s_{n-1}) + p''(s_{n-1})\eta_{n-1}}{p'(s_{n-1})} = 1 - M_{n-1}\eta_{n-1} = 1 - h_{n-1}, \qquad (1.31)$$

so that

$$\eta_n = \frac{1}{2}\frac{M_{n-1}\eta_{n-1}^2}{1 - h_{n-1}} = \frac{h_{n-1}\eta_{n-1}}{2(1 - h_{n-1})}. \qquad (1.32)$$

Similarly, using (1.31), we have

$$M_n = -\frac{2M}{p'(s_n)} = -\frac{2M}{p'(s_{n-1})}\frac{p'(s_{n-1})}{p'(s_n)} = \frac{M_{n-1}}{1 - h_{n-1}}.$$

Hence,

$$h_n = M_n\eta_n = \frac{M_{n-1}h_{n-1}\eta_{n-1}}{2(1 - h_{n-1})^2} = \frac{1}{2}\left(\frac{h_{n-1}}{1 - h_{n-1}}\right)^2. \qquad (1.33)$$

Using now $h_0 = h \le \frac{1}{2}$, from (1.32) and (1.33), we can prove by mathematical induction on n that

$$\eta_n \le h_{n-1}\eta_{n-1}, \qquad h_n \le 2h_{n-1}^2 \le \frac{1}{2}(2h_0)^{2^n} \le \frac{1}{2}, \qquad n \in \mathbb{N},$$

and, as a consequence,

$$\eta_n \le h_{n-1}\eta_{n-1} \le \cdots \le h_{n-1}h_{n-2}\cdots h_0\eta_0 \le \frac{1}{2^n}(2h_0)^{2^n-1}\eta_0.$$

After that, using (1.30) and the last fact, we deduce

$$
\begin{aligned}
s_{n+m} - s_n &= \sum_{j=0}^{m-1} (s_{n+j+1} - s_{n+j}) \\
&= \sum_{j=0}^{m-1} \eta_{n+j} \\
&\leq \sum_{j=0}^{m-1} \frac{1}{2^{n+j}} (2h_0)^{2^{n+j}-1} \eta_0 \\
&\leq \frac{1}{2^n} (2h_0)^{2^n-1} \eta_0 \left(1 + \frac{1}{2}(2h_0)^{2^n} + \frac{1}{2^2}(2h_0)^{2^{n+1}} + \cdots + \frac{1}{2^{m-1}}(2h_0)^{2^m}\right)
\end{aligned}
$$

where $\eta_0 = \eta$ and $h_0 = h$. In addition, by letting $m \to +\infty$ in the last, it follows

$$
s^* - s_n \leq \frac{2}{2^n}(2h_0)^{2^n-1} \eta_0 = \frac{1}{2^{n-1}}(2h_0)^{2^n-1} \eta_0,
$$

where $\eta_0 = \eta$ and $h_0 = h$. In view of (1.27), this yields inequality (1.2) we were seeking. The proof is complete. ∎

This approach rests essentially on the requirement that the majorizing method has the same form as the underlying method. A closer study of this Kantorovich approach reveals that the underlying it is a very simple principle (see [73]). For this principle to be useful, a mechanism is needed to obtain a scalar sequence $\{t_n\}$, that verifies (1.13), for $\{x_n\}$ given.

Remark 1.24. Ortega gives a short and elegant proof of the Newton-Kantorovich theorem using majorizing real sequences in [63], where upper bounds for the errors of Newton's method are known and taught. Moreover, after the appearance of the Newton-Kantorovich theorem, sharp upper and lower bounds for the errors of the method have been derived by many authors under the assumptions of the Newton-Kantorovich theorem or under closely related ones. In particular, in [78], Tapia adds to Ortega's proof a very short and simple derivation of the optimum error estimates. Dennis also derives these optimum error estimates in [15]. Best possible upper and lower bounds for the error are established under the hypotheses of the Newton-Kantorovich theorem by Gragg and Tapia in [36]. In the derivation of one of the optimal upper bounds, Gragg and Tapia resorted to the original Kantorovich recurrence relations. The establishment of these relations is somewhat lengthy and involved. In [60], Miel incorporates in Ortega's proof a derivation of the Gragg-Tapia optimal upper bounds. Yamamoto in [80], and a series of papers, compare the upper and lower bounds obtained by several authors. Yamamoto also gives an interesting method for finding sharp a posteriori error bounds for Newton's method under the Kantorovich assumptions, where a comparison of the best known bounds is also provided, is the best source for error estimates in the theory of the Newton method. In [83], Yamamoto traces historical developments in convergence theory and error estimates for Newton (and Newton-like) methods based on the Newton-Kantorovich theorem.

Remark 1.25. The "method of majorizing sequences" is also a powerful tool that have been applied to establish convergence theorems for variants of Newton's method, so that the concept of a majorizing sequence and estimates of type (1.25) have been extended [73] to give a convergence theory for a large class of iterative methods.

1.1.4   New approach: a "majorant function" from an initial value problem

As we have just seen, the scalar function involved plays a key role in the construction of a majorizing real sequence. For this reason, this study is based on the construction and existence of this scalar function to obtain majorizing real sequences. Thus, we define the following concept of majorant function, highlighting that it is different from that defined by Kantorovich and, from now on, we use it throughout this textbook.

Definition 1.26. *Let $f \in C^1([t_0, +\infty))$ be a real function and Newton's sequence*

$$t_{n+1} = N_f(t_n) = t_n - \frac{f(t_n)}{f'(t_n)}, \quad n \geq 0, \quad \text{with } t_0 \text{ given.} \tag{1.34}$$

If (1.34) is a majorizing sequence of Newton's sequence $\{x_{n+1} = N_F(x_n)\}_{n\geq 0}$ given in (1.1) and defined in the Banach space X, then the real function f is called "majorant function" of the operator F.

Observe that the concept of majorant function given by Kantorovich is applied to operators $\mathfrak{S}$ defined on a Banach space X with values in X ($\mathfrak{S} : X \longrightarrow X$), while the concept of majorant function given in Definition 1.26 is applied to operators F defined on a Banach space X with values in a Banach space Y ($F : X \longrightarrow Y$).

Note that Kantorovich's theory applied to the modified Newton method would have a close relationship with our development if it is applied to this method. However, in the case of Newton's method, we have that if the scalar function N_f majorizes the operator N_F in the sense of Kantorovich, then f generates a majorizing real sequence and, therefore, we have that the function f majorizes the operator F in the sense of Definition 1.26. However, on the contrary, this fact is not true in general. The concept of majorant function given in Definition 1.26 is more general than that given by Kantorovich.

From the modification of the majorant principle of Kantorovich presented before, the first aim is to construct a majorant function f of the operator F. Next, we prove that Newton's sequence (1.34) is convergent and we see finally, as a consequence, that Newton's sequence (1.1) is convergent.

Existence of a majorant function

Once the concept of majorant function is defined, we give first a result that provides conditions that a real function f must satisfy to majorize the operator F.

Theorem 1.27. (General semilocal convergence) *Let $F : \Omega \subseteq X \longrightarrow Y$ be a twice continuously Fréchet differentiable operator defined on a non-empty open convex domain Ω of a Banach space X with values in a Banach space Y. Suppose that there exists $f \in C^2([t_0, +\infty))$ such that the following conditions are satisfied:*

(a) There exists $\Gamma_0 = [F'(x_0)]^{-1} \in \mathcal{L}(Y, X)$, for some $x_0 \in \Omega$, such that $\|\Gamma_0\| \leq -\dfrac{1}{f'(t_0)}$

and $\|\Gamma_0 F(x_0)\| \leq -\dfrac{f(t_0)}{f'(t_0)}$.

(b) $\|F''(x)\| \leq f''(t)$, for $\|x - x_0\| \leq t - t_0$, $x \in \Omega$ and $t \in [t_0, +\infty)$.

If the equation $f(t) = 0$ has only one solution t^* in $[t_0, +\infty)$ and $B(x_0, t^* - t_0) \subset \Omega$, then Newton's method defined in (1.1) and starting at x_0 converges to a solution x^* of $F(x) = 0$. Moreover,

$$\|x^* - x_n\| \leq t^* - t_n, \quad n = 0, 1, 2, \ldots,$$

where t_n is defined in (1.34).

Proof. We prove first that (1.34) is a nondecreasing sequence and convergent to t^*. To this end, we note that sequence (1.34) is monotone and bounded.

As $N_f'(t) \geq 0$ in $[t_0, t^*)$, N_f is nondecreasing in $[t_0, t^*)$ and it follows that t_n is well-defined, for all $n \in \mathbb{N}$, and $t_n \leq t^*$, for all $n = 0, 1, 2, \ldots$, where t^* denotes the solution of $f(t) = 0$ whose existence is postulated in the theorem. In fact, when $n = 0$, the inequality $t_0 \leq t^*$ is obvious; and if it has already been proved for $n = j$, then $t_j \leq t^*$, since N_f is nondecreasing, and implies $N_f(t_j) \leq N_f(t^*)$; that is, $t_{j+1} \leq t^*$. By mathematical induction on n, we have $t_n \leq t^*$, for all $n = 0, 1, 2, \ldots$

Using the fact that N_f is nondecreasing, we can also prove by mathematical induction on n that sequence (1.34) is monotone. For $t_n \leq t_{n+1}$ implies $t_{n+1} = N_f(t_n) \leq N_f(t_{n+1}) = t_{n+2}$ and the inequality $t_0 \leq t_1$ is a consequence of condition (a).

Thus, we have proved that there exists $v = \lim_n t_n$ and, by (1.34) and the continuity of N_f, v is a solution of $f(t) = 0$; moreover, as t^* is the unique solution of $f(t) = 0$ in $[t_0, +\infty)$, then $v = t^*$.

Next, we prove that the elements x_n are well-defined and form a convergent sequence. Firstly, the element x_1 is well-defined, since $x_1 = x_0 - \Gamma_0 F(x_0)$ and there exists Γ_0. Moreover, from

$$\|x_1 - x_0\| \leq \|\Gamma_0 F(x_0)\| \leq -\frac{f(t_0)}{f'(t_0)} = t_1 - t_0 \leq t^* - t_0,$$

we see that $x_1 \in B(x_0, t^* - t_0) \subset \Omega$.

Assume it has been proved that $x_1, x_2, \ldots, x_n \in \Omega$ and

$$\|x_{j+1} - x_j\| \leq t_{j+1} - t_j, \quad \text{for all} \quad j = 0, 1, \ldots, n - 1. \tag{1.35}$$

Now, if we write x and t for corresponding points in $[x_{n-1}, x_n]$ and $[t_{n-1}, t_n]$; that is, if

$$x = x_{n-1} + \tau(x_n - x_{n-1}), \quad t = t_{n-1} + \tau(t_n - t_{n-1}), \quad \tau \in [0, 1],$$

then, in view of (1.35), we have

$$\begin{aligned}
\|x - x_0\| &\leq \tau\|x_n - x_{n-1}\| + \|x_{n-1} - x_{n-2}\| + \cdots + \|x_1 - x_0\| \\
&\leq \tau(t_n - t_{n-1}) + (t_{n-1} - t_{n-2}) + \cdots + (t_1 - t_0) \\
&= t - t_0.
\end{aligned}$$

Hence,

$$\|I - \Gamma_{n-1}F'(x_n)\| = \left\| \int_{x_{n-1}}^{x_n} \Gamma_{n-1}F''(x)\,dx \right\|$$

$$\leq \left\| \int_0^1 \Gamma_{n-1}F''\left(x_{n-1} + \tau(x_n - x_{n-1})\right)(x_n - x_{n-1})\,d\tau \right\|$$

$$\leq \|\Gamma_{n-1}\| \int_0^1 \|F''\left(x_{n-1} + \tau(x_n - x_{n-1})\right)\| \, \|x_n - x_{n-1}\|\,d\tau$$

$$\leq -\frac{1}{f'(t_{n-1})} \int_0^1 f''\left(t_{n-1} + \tau(t_n - t_{n-1})\right)(t_n - t_{n-1})\,d\tau$$

$$\leq -\frac{1}{f'(t_{n-1})} \int_{t_{n-1}}^{t_n} f''(\tau)\,d\tau$$

$$= 1 - \frac{f'(t_n)}{f'(t_{n-1})}$$

$$< 1,$$

since $f''(t) \geq 0$ in $[t_{n-1}, +\infty)$, $f(t_{n-1}) \geq 0$ and $f'(t_n) < 0$. As a consequence, by the Banach lemma on invertible operators, there exists the operator Γ_n and

$$\|\Gamma_n\| \leq \frac{\|\Gamma_{n-1}\|}{1 - \|I - \Gamma_{n-1}F'(x_n)\|} \leq -\frac{1}{f'(t_n)}.$$

After that, by Taylor's series and (1.1), we have

$$F(x_n) = \int_{x_{n-1}}^{x_n} F''(x)(x_n - x)\,dx$$

$$= \int_0^1 F''\left(x_{n-1} + \tau(x_n - x_{n-1})\right)(x_n - x_{n-1})^2(1 - \tau)\,d\tau,$$

so that

$$\|F(x_n)\| \leq \int_0^1 f''\left(t_{n-1} + \tau(t_n - t_{n-1})\right)(t_n - t_{n-1})^2(1 - \tau)\,d\tau$$

$$= \int_{t_{n-1}}^{t_n} f''(\tau)(t_n - \tau)\,d\tau$$

$$= f(t_n).$$

Therefore,

$$\|x_{n+1} - x_n\| \leq \|\Gamma_n\| \|F(x_n)\| \leq -\frac{f(t_n)}{f'(t_n)} = t_{n+1} - t_n,$$

$$\|x_{n+1} - x_0\| \leq \|x_{n+1} - x_n\| + \|x_n - x_0\| \leq t_{n+1} - t_n + t_n - t_0 = t_{n+1} - t_0 \leq t^* - t_0$$

and $x_{n+1} \in B(x_0, t^* - t_0) \subset \Omega$.

Then, by mathematical induction, we have proved that $x_j \in B(x_0, t^* - t_0) \subset \Omega$ and that (1.35) are true for all positive integers j. As a consequence, sequence (1.34) majorizes the sequence $\{x_n\}$ defined in (1.1), so that there exists $\lim_n x_n = x^*$ and $\|x^* - x_n\| \leq t^* - t_n$, for all $n = 0, 1, 2, \ldots$

Now, by the Mean-Value Theorem, we have

$$\|F'(x_n) - F'(x_0)\| \leq \|x_n - x_0\| \sup_{0<\theta<1} \|F''(x_0 + \theta(x_n - x_0))\| \leq (t^* - t_0)\left(\max_{[t_0,t_n]}\{f''(t)\}\right)$$

and we infer, from (1.28), $F(x_n) + F'(x_n)(x_{n+1} - x_n) = 0$, $n \geq 0$, that the sequence $\{\|F'(x_n)\|\}$ is bounded. In addition, taking limits in (1.28) and bearing in mind that F is continuous, we obtain $F(x^*) = 0$, so that x^* is a solution of $F(x) = 0$. ∎

Note that Theorem 1.27 has already been proved from the majorant principle of Kantorovich (see Theorem 1.9) and, now, it has been proved from the modification of the majorant principle given previously which takes into account the concept of majorizing sequence. Observe that conditions (a)-(b) of Theorem 1.9 are conditions (a)-(b) of Theorem 1.27 extended to the domain Ω.

Remark 1.28. Note that conditions (a)-(b) of Theorem 1.27 guarantee that f is a majorant function of the operator F, while the condition required to the equation $f(t) = 0$ guarantees the convergence of the sequence $\{t_{n+1} = N_f(t_n)\}_{n\geq0}$ given in (1.34) and, as a consequence, the convergence of the sequence $\{x_{n+1} = N_F(x_n)\}_{n\geq0}$, given in (1.1), in the Banach space X.

To obtain the majorizing sequence given in (1.34), we have seen that Kantorovich considers a problem of interpolation fitting, where the function f is the second-degree polynomial given in (1.23), since it is the simplest scalar function with bounded second derivative and the coefficient of the polynomial fixed by means of conditions (H1)-(H2)-(H3). But, on the other hand, from Theorem 1.27, that guarantees the existence of a majorant function f, we see that polynomial (1.23) can be obtained otherwise, without interpolation fitting, by solving an initial value problem. Indeed, from (H3) and condition (b) of Theorem 1.27, we observe that

$$\|F''(x)\| \leq M = f''(t) \quad \text{for} \quad \|x - x_0\| \leq t - t_0,$$

so that, once $t_0 > 0$ is fixed, we solve the initial value problem

$$\begin{cases} y''(t) - M = 0, \\ y(t_0) = \dfrac{\eta}{\beta}, \quad y'(t_0) = -\dfrac{1}{\beta}, \end{cases}$$

and obtain the quadratic polynomial given in (1.21) as the unique solution, since, by condition (a) of Theorem 1.27, it follows $-\frac{1}{f'(t_0)} = \beta$ and $-\frac{f(t_0)}{f'(t_0)} = \eta$.

In addition, by solving the previous initial value problem, we can find a majorant function f of the operator F in the sense of Definition 1.26.

Theorem 1.29. *For any nonnegative real numbers M, $\beta \neq 0$ and η, there exists only one solution $f(t)$ of the last initial value problem in $[0, +\infty)$; that is: quadratic polynomial (1.21). Moreover, $f(t)$ is a majorant function of the operator F that satisfies conditions (a)-(b) of Theorem 1.27.*

Finally, from condition (H4), it follows that $f(t)$ has two positive real roots t^* and t^{**} such that $t^* \leq t^{**}$ and, as a consequence, the Newton-Kantorovich theorem is deduced from Theorem 1.29, since Newton's method applied to the majorant function $f(t)$ provides a convergent sequence.

The classic conditions of Kantorovich

Note that Kantorovich's conditions (H1)-(H2)-(H3)-(H4) can be classified into three types: two conditions on the starting point x_0 (conditions (H1)-(H2)), a condition on the operator F (condition (H3)) and a combination condition (condition (H4)). Taking into account this classification, we rewrite conditions (H1)-(H2)-(H3)-(H4) in the following way:

(A1) There exist $\Gamma_0 = [F'(x_0)]^{-1} \in \mathcal{L}(Y, X)$, for some $x_0 \in \Omega$, with $\|\Gamma_0\| \leq \beta$ and $\|\Gamma_0 F(x_0)\| \leq \eta$.

(A2) There exists a constant $M \geq 0$ such that $\|F''(x)\| \leq M$ for $x \in \Omega$.

(A3) $p(\epsilon) \leq 0$, where p is the quadratical polynomial defined in (1.23) and ϵ is the unique positive solution of $p'(s) = 0$, and $B(x_0, s^*) \subset \Omega$, where $s^* = \frac{1-\sqrt{1-2h}}{h} \eta$, with $h = M\beta\eta$, is the smallest positive solution of $p(s) = 0$.

Observe that conditions (H4) and (A3) are equivalent. From now on, we use this way to mention the classic conditions of Kantorovich. Taking into account this, the Newton-Kantorovich theorem can be then written as follows.

Theorem 1.30. *Let $F : \Omega \subseteq X \longrightarrow Y$ be a twice continuously Fréchet differentiable operator defined on a non-empty open convex domain Ω of a Banach space X with values in a Banach space Y. Suppose that conditions (A1)-(A2)-(A3) are satisfied. Then, Newton's sequence defined in (1.1) and starting at x_0 converges to a solution x^* of the equation $F(x) = 0$ and the solution x^* and the iterates x_n belong to $\overline{B(x_0, s^*)}$, for all $n = 0, 1, 2, \ldots$ Moreover, if $h = M\beta\eta < \frac{1}{2}$, the solution x^* is unique in $B(x_0, s^{**}) \cap \Omega$, where $s^{**} = \frac{1+\sqrt{1-2h}}{h} \eta$, and, if $h = \frac{1}{2}$, x^* is unique in $\overline{B(x_0, s^*)}$. Furthermore, we have the error estimates given in (1.2).*

Remark 1.31. The new way of getting polynomial (1.23) in Theorem 1.29 has the advantage of being able to be applied to conditions more general than (A1)-(A2). As a consequence of this, we can extend Kantorovich's theory to conditions more general than (A1)-(A2) by the modification of the majorant principle of Kantorovich above developed, as we see throughout this textbook.

Remark 1.32. Note that condition (A2) of Kantorovich is often difficult to verify because the second derivative is not frequently bounded in the domain of definition of the operator involved, as we see throughout the textbook. In addition, it is not easy to find a domain where condition (A2) is satisfied and contains a solution of the equation that is sought after. To solve the last problem, a common option is to locate previously a solution of the equation to solve in a domain and look for a bound M to satisfy condition (A2) in such domain.

1.2   Applications

We illustrate the results presented in this textbook with applications where nonlinear integral and differentiable problems are involved. As a large number of problems in applied mathematics and engineering are solved by finding the solutions of certain integral equations and boundary value problems, we then pay attention to these two kind of problems in this textbook.

Notice that, in the examples included in the textbook, we use the max-norm for $\mathbb{R}^m$ and $\mathcal{C}([a, b])$ and the norm $\|x\| = \max\{\|x\|_\infty, \|x'\|_\infty, \|x''\|_\infty\}$ for $\mathcal{C}^2([a, b])$, where $\|x\|_\infty =$

$\max_{[a,b]} |x(s)|$ is the norm for $\mathcal{C}([a,b])$, but they are denoted in the same way, $\|\cdot\|$, throughout the entire textbook. We also use, in all the examples included in the textbook, stopping criteria $\|x_n - x_{n-1}\| < 10^{-16}$, $n \in \mathbb{N}$, for Newton's sequence $\{x_n\}$ and $\|x^* - x_n\| < 10^{-16}$, $n \geq 0$, for absolute errors. In addition, we also see that x^* is a good approximation of a solution of the equations involved, $F(x) = 0$, from $\|F(x^*)\| \leq \text{constant} \times 10^{-16}$, so that we include the sequence $\{\|F(x_n)\|\}_{n \geq 0}$ in all the examples.

1.2.1 Integral equations

The first application considered is the approximation of solutions of nonlinear integral equations. In particular, we consider nonlinear integral equations of mixed Hammerstein-type:

$$x(s) = u(s) + \int_a^b \mathcal{K}(s,t)\mathcal{H}(x(t))\,dt, \quad s \in [a,b], \tag{1.36}$$

where $-\infty < a < b < +\infty$, $u \in \mathcal{C}([a,b])$ is a given, the kernel $\mathcal{K}(s,t)$ is a known function in $[a,b] \times [a,b]$, $\mathcal{H}(x)$ is a known function, $\mathcal{H} : \mathcal{C}([a,b]) \longrightarrow \mathcal{C}([a,b])$, and $x \in \mathcal{C}([a,b])$ is the function to determine.

Nonlinear integral equation (1.36) is a particular case of Hammerstein equations of the second kind [67]. The Hammerstein equations have strong physical background and arise from the electro-magnetic fluid dynamics [70]. These equations appeared in the 1930s as general models for the study of semi-linear boundary value problems, where the kernel $\mathcal{K}(s,t)$ typically arises as the Green function of a differential operator [31]. So, these type of equations can be reformulated as a two-point boundary value problem with a certain nonlinear boundary condition [7]. Also multi-dimensional analogues of these equations appear as reformulations of elliptic partial differentiable equations with nonlinear boundary conditions [6]. The Hammerstein equations appear very often in several applications to real world problems [9]. For example, some problems considered in vehicular traffic theory, biology and queuing theory lead to integral equations of this type [14]. Hammerstein equations are also applied in the theory of radiative transfer and the theory of neutron transport as well in the kinetic theory of gases (see [44], among others) and play a very significant role in several applications [12], as for example the dynamic models of chemical reactors [10], which are governed by control equations, justifying then their study and solution [34].

As the Hammerstein equations of form (1.36) cannot be solved exactly, we can use iterative methods to solve them. We then apply Newton's method and use the theoretical significance of the method to draw conclusions about the existence and uniqueness of solution.

Solving equation (1.36) is equivalent to solving the equation $\mathcal{F}(x) = 0$, where $\mathcal{F} : \mathcal{C}([a,b]) \longrightarrow \mathcal{C}([a,b])$ and

$$[\mathcal{F}(x)](s) = x(s) - u(s) - \int_a^b \mathcal{K}(s,t)\mathcal{H}(x(t))\,dt, \quad s \in [a,b]. \tag{1.37}$$

To approximate a solution of the last equation, which in general is nonlinear, by using Newton's method, we take into account how the kernel $\mathcal{K}(s,t)$ is. On the one hand, if $\mathcal{K}(s,t)$ is separabale of the form:

$$\mathcal{K}(s,t) = \sum_{i=1}^n \mathfrak{f}_i(s)\mathfrak{g}_i(t) \quad (a \leq s, t \leq b),$$

where the functions $f_1, f_2, \ldots, f_n, g_1, g_2, \ldots, g_n$ are all continuous on $[a, b]$, we approximate the solutions continuously, without transferring the continuous problem, through a process of discretization, into a discret problem. On the other hand, if $\mathcal{K}(s, t)$ is not separable, we have two possibilities. First of all, if $\mathcal{K}(s, t)$ can be developed in Taylor's series, then the series is truncated, the kernel $\mathcal{K}(s, t)$ can be approximated by a separable kernel and, in this case, we proceed as in the previous case. Secondly, if $\mathcal{K}(s, t)$ is not approximated by a separable kernel, we then use a process of discretization which transforms the original problem into a finite dimensional problem, by means of a known numerical integration formula, and the solutions of the nonlinear system of equations that arise are then approximated by Newton's method.

Application to an integral equation

We present an application to an integral equation where the Newton-Kantorovich theorem is used. So, we consider a nonlinear integral equation of type (1.36), whose solution is equivalent, as we have seen, to solving the equation $\mathcal{F}(x) = 0$, where $\mathcal{F}$ is defined in (1.37).

For operator (1.37), if $\mathcal{H}$ is sufficiently differentiable, we have

$$[\mathcal{F}'(x)y](s) = y(s) - \int_a^b \mathcal{K}(s, t)\mathcal{H}'(x(t))y(t)\, dt,$$

$$[\mathcal{F}''(x)(yz)](s) = -\int_a^b \mathcal{K}(s, t)\mathcal{H}''(x(t))z(t)y(t)\, dt$$

and $\|\mathcal{F}''(x)\| \leq S\|\mathcal{H}''(x)\|$, where $S = \left\|\int_a^b \mathcal{K}(s, t)\,|\,dt\right\|$. Observe then that, in general, condition (A2) of Kantorovich is not satisfied because $\|\mathcal{F}''(x)\|$ is not bounded in $\mathcal{C}([a, b])$ except in certain cases where the funciton $\mathcal{H}''$ is bounded. In general, it is not easy to find a domain $\Omega \subseteq \mathcal{C}([a, b])$ where $\|\mathcal{F}''(x)\|$ is bounded and contains a solution of the equation that is sought after.

To solve the last problem, a common option is to locate previously a solution $x^*(s)$ of equation (1.36) in a domain $\Omega \subseteq \mathcal{C}([a, b])$ and look for a bound for $\|\mathcal{F}''(x)\|$ in Ω (see Remark 1.32). So, a solution $x^*(s)$ of (1.36) must satisfy that

$$\|x^*(s)\| - U - S\|\mathcal{H}(x^*(s))\| \leq 0, \tag{1.38}$$

where $U = \|u(s)\|$. Therefore, from (1.38), we try to find a region $\Omega \subseteq \mathcal{C}([a, b])$ that contains $x^*(s)$.

However, equation (1.38), which $x^*(s)$ must satisfy, can be verified in different regions of the space, on the basis of the number of positive real roots of the scalar equation deduced from (1.38) and given by $\chi(t) = t - U - S\|\mathcal{H}(t)\| = 0$. In particular, equation (1.38) can be satisfied at intervals of the form $[0, r_1]$, $[r_2, r_3]$ or $[r_4, +\infty)$, where r_1, r_2, r_3 and r_4 are positive real roots of the previous scalar equation. Then, previously, we can only locate solutions $x^*(s)$ such that $\|x^*(s)\| \in [0, r_1]$ or $\|x^*(s)\| \in [r_2, r_3]$. In the other case, when $\|x^*(s)\| > r_4$, the problem remains open. Observe that the case where the previous scalar equation does not have positive real roots is analogous to the last one. Other situations for the number of roots are reduced to some of the last ones.

Now, we consider nonlinear integral equations of Hammerstein-type (1.36) with

$$\mathcal{H}(x(t)) = \lambda_1\, x(t)^{2+p} + \lambda_2\, x(t)^q, \quad p \in [0, 1], \quad q \in \mathbb{N}, \quad \lambda_1, \lambda_2 \in \mathbb{R}. \tag{1.39}$$

If we choose $a = 0$, $b = 1$, $u(s) = s^2$, $\mathcal{K}(s,t) = st^{13}$, $\lambda_1 = 1$, $\lambda_2 = \frac{1}{2}$, $p = \frac{1}{3}$ and $q = 2$, then (1.36)-(1.39) is reduced to

$$x(s) = s^2 + s \int_0^1 t^{13} \left(x(t)^{7/3} + \frac{1}{2} x(t)^2 \right) dt, \quad s \in [0,1], \tag{1.40}$$

where x is a solution to determine.

Then, solving equation (1.40) is equivalent to solving $\mathcal{F}(x) = 0$, where $\mathcal{F} : \mathcal{C}([0,1]) \longrightarrow \mathcal{C}([0,1])$ is such that

$$[\mathcal{F}(x)](s) = s^2 - s \int_0^1 t^{13} \left(x(t)^{7/3} + \frac{1}{2} x(t)^2 \right) dt. \tag{1.41}$$

Moreover, for operator (1.41), we have

$$[\mathcal{F}'(x)y](s) = y(s) - s \int_0^1 t^{13} \left(\frac{7}{3} x(t)^{4/3} + x(t) \right) y(t) \, dt,$$

$$[\mathcal{F}''(x)(yz)](s) = -s \int_0^1 t^{13} \left(\frac{28}{9} x(t)^{1/3} + 1 \right) z(t) y(t) \, dt$$

and condition (A2) of the Newton-Kantorovich theorem is not satisfied in general because $\|\mathcal{F}''(x)\|$ is not bounded in $\mathcal{C}([0,1])$ because $\mathcal{H}''(x(s)) = \frac{28}{9} x(s)^{1/3} + 1$ is not bounded, in general, in $\mathcal{C}([0,1])$.

So, from (1.38), it follows that $x^*(s)$ must satisfy that $\|x^*(s)\| \leq \rho_1 = 1.1447\dots$ or $\|x^*(s)\| \geq \rho_2 = 5.0672\dots$, where ρ_1 and ρ_2 are the two real positive roots of the scalar equation $\chi(t) = t - 1 - \frac{1}{14} \left(t^{7/3} + \frac{1}{2} t^2 \right) = 0$ deduced from (1.38). Thus, from the Newton-Kantorovich theorem, we can only approximate one solution $x^*(s)$ by Newton's method, that which satisfies $\|x^*(s)\| \in [0, \rho_1]$, since we can consider the domain

$$\Omega = \{x(s) \in \mathcal{C}([0,1]) : \|x(s)\| < \rho, \ s \in [0,1]\},$$

with $\rho \in (\rho_1, \rho_2)$, as domain for the operator $\mathcal{F}$, where $\|\mathcal{F}''(x)\|$ is bounded, and takes $x_0(s) \in \Omega$ as starting point. For this, we take, for example, $\rho = 3$ and choose, as it is usually done ([21, 26, 27]), the starting point $x_0(s) = s^2$.

As the kernel of (1.40), $s t^{13}$, is separable, we can then determine the corresponding operator $[\mathcal{F}'(x)]^{-1}$. For this, we write $[\mathcal{F}'(x)y](s) = \zeta(s)$, so that, if there exists $[\mathcal{F}'(x)]^{-1}$, we have

$$[\mathcal{F}'(x)]^{-1}\zeta(s) = y(s) = \zeta(s) + s \int_0^1 t^{13} \left(\frac{7}{3} x(t)^{4/3} + x(t) \right) y(t) \, dt.$$

If we now denote $\int_0^1 t^{13} \left(\frac{7}{3} x(t)^{4/3} + x(t) \right) y(t) \, dt = \mathcal{I}$, multiply next-to-last equality by $s^{13} \left(\frac{7}{3} x(s)^{4/3} + x(s) \right)$ and integrate it between 0 and 1, we obtain

$$\mathcal{I} = \frac{\int_0^1 s^{13} \left(\frac{7}{3} x(s)^{4/3} + x(s) \right) \zeta(s) \, ds}{1 - \int_0^1 s^{14} \left(\frac{7}{3} x(s)^{4/3} + x(s) \right) ds}$$

provided that

$$\int_0^1 s^{14} \left(\frac{7}{3} x(s)^{4/3} + x(s) \right) ds \neq 1. \tag{1.42}$$

Therefore,

$$y(s) = [\mathcal{F}'(x)]^{-1}\zeta(s) = \zeta(s) + s \frac{\int_0^1 t^{13} \left(\frac{7}{3} x(t)^{4/3} + x(t) \right) \zeta(t) \, dt}{1 - \int_0^1 t^{14} \left(\frac{7}{3} x(t)^{4/3} + x(t) \right) dt}.$$

As a consequence, $\|[\mathcal{F}'(x_0)]^{-1}\| \leq 1.2502\ldots = \beta$ and $\|[\mathcal{F}'(x_0)]^{-1}\mathcal{F}(x_0)\| \leq 0.1017\ldots = \eta$.

Since $\|\mathcal{F}''(x)\| \leq (0.3919\ldots) = M$ and $M\beta\eta = 0.0498\ldots \leq \frac{1}{2}$. Hence, the Newton-Kantorovich theorem is applied, so that $p(s) = (0.0813\ldots) - (0.7998\ldots)s + (0.1959\ldots)s^2$, $s^* = 0.1043\ldots$, and $B(x_0, s^*) \subset \Omega = \{x(s) \in \mathcal{C}([0,1]) : \|x(s)\| < 3, \ s \in [0,1]\}$. As a consequence, Newton's method can be used to approximate a solution $x^*(s)$ in Ω.

Taking into account the last and $s^{**} = 3.9770\ldots$, the Newton-Kantorovich theorem says that the domains of existence and uniqueness of solution are respectively

$$\{\nu \in \Omega : \|\nu(s) - x_0(s)\| \leq 0.1043\ldots\} \quad \text{and} \quad \{\nu \in \Omega : \|\nu(s) - x_0(s)\| < 3.9770\ldots\}.$$

Note that condition (1.42) is satisfied in $B(x_0, s^*)$ for all $x(s)$, so that the sequence $\{x_n(s)\}$, given by Newton's method, is well-defined.

After that, we apply Newton's method for approximating a solution with the features mentioned above. The direct application of Newton's method is

$$x_{n+1}(s) = x_n(s) - [\mathcal{F}'(x_n)]^{-1}[\mathcal{F}(x_n)](s) = s^2 + s\,\frac{\mathcal{A}_n - \mathcal{B}_n + \mathcal{E}_n}{1 - \mathcal{D}_n},$$

where

$$\mathcal{A}_n = \int_0^1 t^{13}\left(x_n(t)^{7/3} + \frac{1}{2}x_n(t)^2\right)dt, \qquad \mathcal{B}_n = \int_0^1 t^{13}\left(\frac{7}{3}x_n(t)^{4/3} + x_n(t)\right)x_n(t)\,dt,$$

$$\mathcal{D}_n = \int_0^1 t^{14}\left(\frac{7}{3}x_n(t)^{4/3} + x_n(t)\right)dt, \qquad \mathcal{E}_n = \int_0^1 t^{15}\left(\frac{7}{3}x_n(t)^{4/3} + x_n(t)\right)dt,$$

and $x^*(s) = (0.1021\ldots)s + s^2$ is the approximated solution obtained after four iterations and shown in Table 1.1. In Table 1.1, we can also see the errors $\|x^*(s) - x_n(s)\|$ and the sequence $\|[\mathcal{F}(x_n)](s)\|$. From the last, we notice that $x^*(s)$ is a good approximation of the solution of (1.40).

Moreover, see Figure 1.1, the approximated solution $x^*(s)$ lies within the existence domain of solution obtained above and observe that $\|x^*(s)\| = 1.1021\ldots \leq \rho_1 = 1.1447\ldots$

n	$x_n(s)$	$\|x^*(s) - x_n(s)\|$	$\|[\mathcal{F}(x_n)](s)\|$
0	s^2	$1.0216\ldots \times 10^{-1}$	$8.1349\ldots \times 10^{-2}$
1	$(0.10054270908180\ldots)s + s^2$	$1.6214\ldots \times 10^{-3}$	$1.2704\ldots \times 10^{-3}$
2	$(0.10216370647960\ldots)s + s^2$	$4.2876\ldots \times 10^{-7}$	$3.3585\ldots \times 10^{-7}$
3	$(0.10216413524012\ldots)s + s^2$	$3.0004\ldots \times 10^{-14}$	$2.3503\ldots \times 10^{-14}$
4	$(0.10216413524015\ldots)s + s^2$		

Table 1.1: Approximated solution $x^*(s)$ of (1.40), absolute errors and $\{\|[\mathcal{F}(x_n)](s)\|\}$

1.2.2 Boundary value problems

The second application that we pay attention is the approximation of solutions of some boundary value problems of second order, which are in general described by the equation

$$\frac{d^2x(s)}{ds^2} = \Phi\left(s, x(s), \frac{dx(s)}{ds}\right)$$

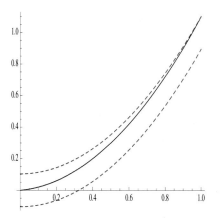

Figure 1.1: Graph (the solid line) of the approximated solution $x^*(s)$ of (1.40)

and the boundary conditions

$$x(a) = A, \qquad x(b) = B,$$

where $\Phi : [a, b] \times \mathbb{R}^2 \longrightarrow \mathbb{R}$.

Boundary value problems of second order usually appear in all branches of experimental sciences. For example: in physics, Newton's laws and many others are expressed as a problem of this type; in biology, they appear in population dynamics modes; in chemistry, they arise in the evaluation of concentrations of different reagents during a reaction, etc. As in the majority of nonlinear problems, the exact solution is not known, so that the solution is numerically approximated. One of the methods used is the method of finite differences, that transforms the differential problem into a system of nonlinear equations, which usually is solved by Newton's method.

It is well known that energy is dissipated in the action of any real dynamical system, usually through some form of friction. However, in certain situations this dissipation is so slow that it can be neglected over relatively short periods of time. In such cases we assume the law of conservation of energy, namely, that the sum of the kinetic energy and the potential energy is constant. A system of this kind is said to be conservative.

If φ and ψ are arbitrary functions with the property that $\varphi(0) = 0$ and $\psi(0) = 0$, the general equation

$$\mathrm{m}\frac{d^2x(s)}{ds^2} + \psi\left(\frac{dx(s)}{ds}\right) + \varphi(x(s)) = 0 \tag{1.43}$$

can be interpreted as the equation of motion of a mass m under the action of a restoring force $-\varphi(x(s))$ and a damping force $-\psi(dx/ds)$. In general, these forces are nonlinear and equation (1.43) can be regarded as the basic equation of nonlinear mechanics. In this textbook we consider the special case of a nonlinear conservative system described by the equation

$$\mathrm{m}\frac{d^2x(s)}{ds^2} + \varphi(x(s)) = 0,$$

in which the damping force is zero and there is consequently no dissipation of energy. Extensive discussions of (1.43), with applications to a variety of physical problems, can be found in classical references [1] and [77].

More specifically, we study the existence of a unique solution for a special case of the nonlinear conservative system described by the equation

$$\frac{d^2 x(s)}{ds^2} + \phi(x(s)) = 0 \tag{1.44}$$

with the boundary conditions

$$x(0) = x(1) = 0, \tag{1.45}$$

where $\phi(x)$ is sufficiently derivable.

We note first that solving boundary value problem (1.44)-(1.45) is equivalent to solving a Fredholm integral equation of the form [68]:

$$x(s) = - \int_0^1 G(s,t) \phi(x(t)) \, dt, \tag{1.46}$$

where the kernel G is the Green function in $[0,1] \times [0,1]$:

$$G(s,t) = \begin{cases} (1-s)t, & t \le s, \\ s(1-t), & s \le t. \end{cases}$$

Second, we can consider the second-order differential equation as an operator $\mathfrak{F} : \mathcal{C}^2([0,1]) \longrightarrow \mathcal{C}([0,1])$ such that

$$[\mathfrak{F}(x)](s) = \frac{d^2 x(s)}{ds^2} + \phi(x(s))$$

and provided that $\phi \in \mathcal{C}^2(\mathbb{R})$.

Another important class of problems for ordinary differential equations is to find a solution $x \in \mathcal{C}([a,b])$ of the second-order differential equation

$$\frac{d^2 x(s)}{ds^2} + \Psi(s, x(s)) = 0 \tag{1.47}$$

which satisfies the boundary conditions

$$x(a) = A, \qquad x(b) = B, \tag{1.48}$$

where $\Psi(s,x)$ is sufficiently derivable in both arguments ([72]).

In this case, solving the previous boundary value problem is equivalent to solving a Fredholm integral equation of the form:

$$x(s) = - \int_a^b G(s,t) \Psi(t, x(t)) \, dt + \frac{B-A}{b-a} s + \frac{bA - Ba}{b-a}, \tag{1.49}$$

where the kernel G is the Green function in $[a,b] \times [a,b]$:

$$G(s,t) = \begin{cases} \dfrac{(b-s)(t-a)}{b-a}, & t \le s, \\ \dfrac{(s-a)(b-t)}{b-a}, & s \le t. \end{cases} \tag{1.50}$$

In addition, we can consider the previous second-order differential equation as an operator $\mathfrak{F} : \mathcal{C}^2([a,b]) \longrightarrow \mathcal{C}([a,b])$ such that

$$[\mathfrak{F}(x)](s) = \frac{d^2 x(s)}{ds^2} + \Psi(s, x(s)).$$

We then apply Newton's method for approximating a solution of the equation $\mathfrak{F}(x) = 0$. For this, a process of discretization is usually used that leads to a system of nonlinear equations that is solved by Newton's method.

Application to a boundary value problem

Now, by using the Newton-Kantorovich theorem, we give an application of Newton's method to a boundary value problem of form (1.44)-(1.45).

For the direct numerical solution of problem (1.44)-(1.45), we introduce the points $t_j = jh$, $j = 0, 1, \ldots, m+1$, where $h = \frac{1}{m+1}$ and m is an appropriate integer. A scheme is then designed for the determination of numbers x_j, it is hoped, approximate the values $x(t_j)$ of the true solution at the points t_j. A standard approximation for the second derivative at these points is

$$x_j'' \approx \frac{x_{j-1} - 2x_j + x_{j+1}}{h^2}, \quad j = 1, 2, \ldots, m.$$

A natural way to obtain such a scheme is to demand that the x_j satisfy at each interior mesh point t_j the difference equation

$$x_{j-1} - 2x_j + x_{j+1} + h^2 \phi(x_j) = 0. \tag{1.51}$$

Since x_0 and x_{m+1} are determined by the boundary conditions, the unknowns are $x_1, x_2, \ldots, x_m$.

A further discussion is simplified by the use of matrix and vector notation. Introducing the vectors

$$\mathbf{x} = \begin{pmatrix} x_1 \\ x_2 \\ \vdots \\ x_m \end{pmatrix}, \quad v_{\mathbf{x}} = \begin{pmatrix} \phi(x_1) \\ \phi(x_2) \\ \vdots \\ \phi(x_m) \end{pmatrix}$$

and the matrix

$$A = \begin{pmatrix} -2 & 1 & 0 & \cdots & 0 \\ 1 & -2 & 1 & \cdots & 0 \\ 0 & 1 & -2 & \cdots & 0 \\ \vdots & \vdots & \vdots & \ddots & \vdots \\ 0 & 0 & 0 & \cdots & -2 \end{pmatrix},$$

the system of equations, arising from demanding that (1.51) holds for $j = 1, 2, \ldots, m$, can be written compactly in the form

$$\mathbb{F}(\mathbf{x}) \equiv A\mathbf{x} + h^2 v_{\mathbf{x}} = 0 \tag{1.52}$$

with $\mathbb{F} : \Lambda \subset \mathbb{R}^m \longrightarrow \mathbb{R}^m$.

If the function $\phi(u)$ is not linear in u, we cannot hope to solve system (1.52) by algebraic methods. Some iterative procedure must be then resorted to and we use Newton's method for this purpose.

The steady temperature distribution is known in a homogeneous rod of length 1 in which, as a consequence of a chemical reaction or some such heat-producing process, heat is generated at a rate $\phi(x(s))$ per unit time per unit length, $\phi(x(s))$ being a given function of the excess temperature x of the rod over the temperature of the surroundings. If the ends of the rod, $t = 0$ and $t = 1$, are kept at given temperatures, we are to solve boundary value problem (1.44)-(1.45), measured along the axis of the rod. For an example, we choose the following polynomial law

$$\phi(x(s)) = 3 + x(s) + 2x(s)^2 + x(s)^3 \tag{1.53}$$

for the heat generation.

According to the above-mentioned, $\mathbf{x} = (x_1, x_2, \ldots, x_m)^T$ and $v_{\mathbf{x}} = (3 + x_1 + 2x_1^2 + x_1^3, 3 + x_2 + 2x_2^2 + x_2^3, \ldots, 3 + x_m + 2x_m^2 + x_m^3)^T$. In addition, the first derivative of the function $\mathbb{F}$ defined in (1.52) is given by

$$\mathbb{F}'(\mathbf{x}) = A + h^2 D(d_{\mathbf{x}}),$$

where $d_{\mathbf{x}} = (1 + 4x_1 + 3x_1^2, 1 + 4x_2 + 3x_2^2, \ldots, 1 + 4x_m + 3x_m^2)^T$ and $D(d_{\mathbf{x}}) = \text{diag}\{1 + 4x_1 + 3x_1^2, 1 + 4x_2 + 3x_2^2, \ldots, 1 + 4x_m + 3x_m^2\}$. Moreover,

$$\mathbb{F}''(\mathbf{x})\mathbf{y}\,\mathbf{z} = (y_1, y_2, \ldots, y_m)\mathbb{F}''(\mathbf{x})(z_1, z_2, \ldots, z_m),$$

where $\mathbf{y} = (y_1, y_2, \ldots, y_m)^T$ and $\mathbf{z} = (z_1, z_2, \ldots, z_m)^T$, so that

$$\mathbb{F}''(\mathbf{x})\mathbf{y}\,\mathbf{z} = h^2\left((4 + 6x_1)y_1z_1, (4 + 6x_2)y_2z_2, \ldots, (4 + 6x_m)y_mz_m\right)^T.$$

Next, we study the application of the Newton-Kantorovich theorem to this problem. As

$$\|\mathbb{F}''(\mathbf{x})\| = \sup_{\substack{\|\mathbf{y}\|=1 \\ \|\mathbf{z}\|=1}} \|\mathbb{F}''(\mathbf{x})\mathbf{y}\,\mathbf{z}\|, \tag{1.54}$$

and

$$\|\mathbb{F}''(\mathbf{x})\mathbf{y}\,\mathbf{z}\| \leq h^2 \left\|\begin{pmatrix} (4 + 6x_1)y_1z_1 \\ (4 + 6x_2)y_2z_2 \\ \vdots \\ (4 + 6x_m)y_mz_m \end{pmatrix}\right\| \leq h^2(4 + 6\|\mathbf{x}\|)\|\mathbf{y}\|\|\mathbf{z}\|,$$

we observe that $\|\mathbb{F}''(\mathbf{x})\|$ is not bounded in general, since the function deduced from the last expression and given by $\wp(t) = 4 + 6t$ is nondecreasing. Therefore, condition (A2) of Kantorovich is not satisfied, so that we cannot apply the Newton-Kantorovich theorem.

As we have indicated previously in Remark 1.32, to solve the last difficulty and apply the Newton-Kantorovich theorem, we can locate the solution $x^*(s)$ in some domain. For this, we have

$$\|x^*(s)\| - \frac{1}{8}\left(3 + \|x^*(s)\| + 2\|x^*(s)\|^2 + \|x^*(s)\|^3\right) \leq 0,$$

where $\frac{1}{8} = \max_{[0,1]} \int_0^1 |G(s,t)|\,dt$, so that $\|x^*(s)\| \in [0, \rho_1] \cup [\rho_2, +\infty]$, where $\rho_1 = 0.5301\ldots$ and $\rho_2 = 1.4291\ldots$ are the two positive real roots of the scalar equation deduced from the last expression and given by $\chi(t) = t - \frac{1}{8}(3 + t + 2t^2 + t^3) = 0$.

By the Newton-Kantorovich theorem, we could only guarantee the semilocal convergence of Newton's method to a solution $x^*(s)$ such that $\|x^*(s)\| \in [0, \rho_1]$. For this, we can consider for the operator F the domain

$$\Omega = \left\{x(s) \in \mathcal{C}^2([0,1]) : \|x(s)\| < \frac{3}{4}, \ s \in [0,1]\right\},$$

since $\rho_1 < \frac{3}{4} < \rho_2$.

In view of what the domain Ω is for equation (1.44), we then consider (1.52) with $\mathbb{F} : \Lambda \subset \mathbb{R}^m \longrightarrow \mathbb{R}^m$ and $\Lambda = \left\{\mathbf{x} \in \mathbb{R}^m : \|\mathbf{x}\| < \frac{3}{4}\right\}$.

If we choose $m = 8$ and the starting point $\mathbf{x_0} = (0, 0, \ldots, 0)^T$, we see that conditions of the Newton-Kantorovich theorem are satisfied, since

$$\|\mathbb{F}''(\mathbf{x})\| \leq \frac{17}{162} = M, \quad \beta = 11.1694\ldots, \quad \eta = 0.4136\ldots \quad \text{and} \quad M\beta\eta = 0.4848\ldots < \frac{1}{2}.$$

In addition, Kantorovich's polynomial is $p(s) = \frac{17}{324}s^2 - (0.0895\ldots)s + (0.0370\ldots)$, the two real zeros of p are $s^* = 0.7047\ldots$, $s^{**} = 1.0015\ldots$ and $B(\mathbf{x_0}, s^*) \subseteq \Lambda = B\left(0, \frac{3}{4}\right)$. Therefore, the domains of existence and uniqueness of solution are respectively

$$\{\mathbf{v} \in \mathbb{R}^8 : \|\mathbf{v}\| \le 0.7047\ldots\} \quad \text{and} \quad \{\mathbf{v} \in \mathbb{R}^8 : \|\mathbf{v}\| < 1.0015\ldots\} \cap \Lambda = \Lambda$$

and Newton's method converges to the solution $\mathbf{x}^* = (x_1^*, x_2^*, \ldots, x_8^*)^T$ shown in Table 1.2 after five iterations. Observe that $\|\mathbf{x}^*\| = 0.4685\ldots \le \frac{3}{4}$.

i	x_i^*	i	x_i^*
1	0.18110836...	5	0.46853793...
2	0.32206056...	6	0.41902620...
3	0.41902620...	7	0.32206056...
4	0.46853793...	8	0.18110836...

Table 1.2: Numerical solution $\mathbf{x}^*$ of (1.52) with $\phi(u)$ defined in (1.53)

In Table 1.3 we show the errors $\|\mathbf{x}^* - \mathbf{x_n}\|$ and the sequence $\{\|\mathbb{F}(\mathbf{x_n})\|\}$. From the last sequence, we can deduce that the vector shown in Table 1.2 is a good approximation of the solution of system (1.52)-(1.53) with $m = 8$.

n	$\|\mathbf{x}^* - \mathbf{x_n}\|$	$\|\mathbb{F}(\mathbf{x_n})\|$
0	$4.6853\ldots \times 10^{-1}$	$3.7037\ldots \times 10^{-2}$
1	$5.4855\ldots \times 10^{-2}$	$5.0995\ldots \times 10^{-3}$
2	$1.1968\ldots \times 10^{-3}$	$1.1711\ldots \times 10^{-5}$
3	$6.0715\ldots \times 10^{-7}$	$6.0118\ldots \times 10^{-8}$
4	$1.5627\ldots \times 10^{-13}$	$1.5499\ldots \times 10^{-14}$

Table 1.3: Absolute errors and $\{\|\mathbb{F}(\mathbf{x_n})\|\}$

By interpolating the values of Table 1.2 and taking into account (1.45), we obtain the solution drawn in Figure 1.2. Observe also that this interpolated solution lies within the existence domain of solution obtained above.

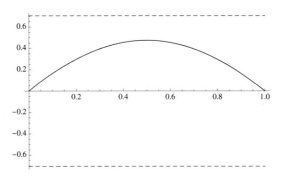

Figure 1.2: Graph (the solid line) of the approximated solution $\mathbf{x}^*$ of (1.44)-(1.45) with $\phi(u)$ defined in (1.53)

Chapter 2

Convergence conditions on the second derivative of the operator

As we have pointed out in the preface to the textbook, our main goal is to obtain semilocal convergence results for Newton's method under different conditions for the operator involved. In Chapter 1, we have seen that the application of the majorant principle of Kantorovich is not easy if condition (A2) is not satisfied. For this, we have introduced a more general definition of majorant function than that given by Kantorovich and, from Theorem 1.27, we have seen the conditions that must be satisfy the scalar function candidate to be a majorant function.

We begin the study in Section 2.1 by focusing our attention on conditions on the second derivative of the operator involved, which is what initially considered Kantorovich, see condition (A2), for his study of Newton's method from the majorant principle. The first study presented is due to Huang [45], who establishes a semilocal convergence result for Newton's method under a Lipschitz condition for the second derivative of the operator involved from the method of majorizing sequences. A first generalization of this condition is given by a condition of Hölder continuity for the second derivative. We go a little further and study the semilocal convergence of Newton's method under an even more general condition, that is named ω-Lipschitz condition for the second derivative and includes as particular cases the previous two. For this, we use the modification of the majorant principle of Kantorovich presented in Section 1.1.4. As it happens with the result of Huang, we see that the semilocal convergence result does not need this condition entirely, but simply that it is centered at the starting point x_0 of Newton's method [26]. Under this ω-Lipschitz condition for the second derivative, we see that we can modify the domain of starting points obtained by Kantorovich for Newton's method and thus guarantee the semilocal convergence of Newton's method from starting points which do not satisfy the classic conditions of Kantorovich, although the conditions imposed are not comparable.

Next, in Section 2.2, we consider a generalization of condition (A2) of Kantorovich by requiring that the second derivative of the operator involved is ω-bounded [25], which is milder than that required by Kantorovich in (A2). The main advantage of this condition with respect to (A2) is that we do not need to locate previously the solution of the equation to solve besides the fact that it is easier to verify than (A2).

© Springer International Publishing AG 2017
J.A. Ezquerro Fernández, M.Á. Hernández Verón, *Newton's Method: an Updated Approach of Kantorovich's Theory*, Frontiers in Mathematics, DOI 10.1007/978-3-319-55976-6_2

2.1 Operators with Lipschitz-type second-derivative

Most of papers studying the semilocal convergence of Newton's method are modifications of the Newton-Kantorovich theorem that relax conditions (A1)-(A2)-(A3), specially condition (A2). But if the condition required to the operator F is milder than (A2), as we can see in [4, 5, 20, 21, 69, 73, 81], then condition (A3) is usually replaced by other condition more restrictive, which necessarily leads to a reduction in the domain of valid starting points for the convergence of Newton's method. In this section, our main aim is not to require milder conditions to the operator F, but stronger, which pursue a modification, not a restriction, of the valid starting points for the convergence of Newton's method, so that this method can start at points from which the Newton-Kantorovich theorem cannot guarantee its semilocal convergence, as well as improving the domains of existence and uniqueness of solution and the a priori error estimates. For this, different conditions are required to the operator F''. Our approach goes through to obtain general semilocal convergence results by using the modification of the majorant principle of Kantorovich given in Section 1.1.3, which allows constructing majorizing real sequences in more general situations than the classic majorant principle of Kantorovich. Finally, from general results, we see others as particular cases ([37, 86]).

2.1.1 Operators with Lipschitz second-derivative

Under conditions (A1)-(A2)-(A3), one can get the error estimates, existence and uniqueness domains of solutions and know whether x_0 is an initial point from which Newton's method converges. But, sometimes, conditions (A1)-(A2)-(A3) fail and Newton's method does not converge starting at x_0, as we can see in the following example.

Example 2.1. We consider a nonlinear integral equation of Hammerstein-type defined in (1.36)-(1.39) with $a = 0$, $b = 1$, $u(s) = \sin s$, $\mathcal{K}(s,t) = s^2 t^7$ and $\mathcal{H}(x(t)) = x(t)^3$; i.e.:

$$x(s) = \sin s + s^2 \int_0^1 t^7 x(t)^3 \, dt, \quad s \in [0,1]. \tag{2.1}$$

Solving equation (2.1) is equivalent to solving $\mathcal{F}(x) = 0$, where $\mathcal{F} : \mathcal{C}([0,1]) \longrightarrow \mathcal{C}([0,1])$ and

$$[\mathcal{F}(x)](s) = x(s) - \sin s - s^2 \int_0^1 t^7 x(t)^3 \, dt.$$

Moreover, we have

$$[\mathcal{F}'(x)y](s) = y(s) - 3s^2 \int_0^1 t^7 x(t)^2 y(t) \, dt,$$

$$[\mathcal{F}''(x)(yz)](s) = -6s^2 \int_0^1 t^7 x(t) z(t) y(t) \, dt,$$

so that condition (A2) of Kantorovich is not satisfied in general because $\|\mathcal{F}''(x)\|$ is not bounded in $\mathcal{C}([0,1])$.

So, if we take norms in (2.1), then we have that $x^*(s)$ must satisfy that $\|x^*(s)\| \le \rho_1 = 0.9479\ldots$ or $\|x^*(s)\| \ge \rho_2 = 2.2326\ldots$, where ρ_1 and ρ_2 are the two real positive roots of the scalar equation $\chi(t) = \frac{t^3}{8} - t + \sin 1 = 0$, which is deduced from (1.38) for equation (2.1). Thus, from the Newton-Kantorovich theorem, we can only approximate one solution $x^*(s)$

by Newton's method, that which satisfies $\|x^*(s)\| \in [0, \rho_1]$, since we can consider the domain $\Omega = \{x(s) \in \mathcal{C}([0,1]) : \|x(s)\| < \rho, s \in [0,1]\}$, with $\rho \in (\rho_1, \rho_2)$, as domain for the operator F, where $\|\mathcal{F}''(x)\|$ is bounded. For this, we choose, for example, $\rho = 2$ and the starting point $x_0(s) = 0$. In addition, $\beta = 1$, $\eta = \sin 1$ and $M = \frac{3}{2}$, so that condition (A3), that is equivalent to $M\beta\eta \leq \frac{1}{2}$ (see Section 1.1.4), is not satisfied, since $M\beta\eta = 1.2622\ldots > \frac{1}{2}$.

In [45], Huang establishes a new semilocal convergence result, that we see later, under which Newton's method, starting at $x_0(s) = 0$ in Example 2.1, converges if the following new conditions are satisfied:

(B1) There exists $\Gamma_0 = [F'(x_0)]^{-1} \in \mathcal{L}(Y, X)$, for some $x_0 \in \Omega$, with $\|\Gamma_0\| \leq \beta$ and $\|\Gamma_0 F(x_0)\| \leq \eta$; moreover, $\|F''(x_0)\| \leq \delta$.

(B2) There exists a constant $K \geq 0$ such that $\|F''(x) - F''(y)\| \leq K\|x - y\|$, for $x, y \in \Omega$.

(B3) $\phi(\alpha) \leq 0$, where ϕ is the function

$$\phi(t) = \frac{K}{6}t^3 + \frac{\delta}{2}t^2 - \frac{t}{\beta} + \frac{\eta}{\beta} \tag{2.2}$$

and α is the unique positive solution of $\phi'(t) = 0$, and $B(x_0, t^*) \subset \Omega$, where t^* is the smallest positive solution of $\phi(t) = 0$.

Note that Huang states in [45] that $\phi(\alpha) \leq 0$ holds provided that one of the two following conditions is satisfied:

$$6\delta^3\beta^3\eta + 9K^2\beta^2\eta^2 + 18K\delta\beta^2\eta - 3\delta^2\beta^2 - 8K\beta \leq 0,$$

$$3K\delta\beta^2 + \delta^3\beta^3 + 3K^2\beta^2\eta \leq \left(\delta^2\beta^2 + 2K\beta\right)^{3/2}.$$

Moreover, in [82], we can also see the following equivalent condition to $\phi(\alpha) \leq 0$:

$$\eta \leq \frac{4K\beta + \delta^2\beta^2 - \delta\beta\sqrt{\delta^2\beta^2 + 2K\beta}}{3K\beta\left(\delta\beta\sqrt{\delta^2\beta^2 + 2K\beta}\right)}.$$

Remark 2.2. As α is the unique positive solution of $\phi'(t) = 0$ and $\phi''(\alpha) > 0$, then α is a minimum of $\phi(t)$ such that $\phi(\alpha) \leq 0$, so that (B3) is a necessary and sufficient condition for the existence of two positive solutions t^* and t^{**} of $\phi(t) = 0$ such that $0 < t^* \leq t^{**}$. Moreover, as ϕ is a nonincreasing convex function in $[0, \alpha]$ such that $\phi(\alpha) \leq 0 < \phi(0)$ and $\phi(0)\phi''(0) > 0$, then these conditions are sufficient to guarantee the semilocal convergence of Newton's nondecreasing real sequence,

$$t_0 = 0, \quad t_{n+1} = N_\phi(t_n) = t_n - \frac{\phi(t_n)}{\phi'(t_n)}, \quad n = 0, 1, 2, \ldots, \tag{2.3}$$

to t^* (see Theorem 1.21). So, from the modification of the majorant principle of Kantorovich given in Section 1.1.3, it is sufficient to prove that the scalar function ϕ majorizes the operator F to guarantee the semilocal convergence of Newton's method in the Banach space X.

Notice that the semilocal convergence conditions required to Newton's method in [45] is not exactly the same that we consider above, but they are equivalent and nothing change the following semilocal convergence theorem established by Huang. The reason for this change is the uniformity sought throughout the present textbook. As a consequence, we have adapted the semilocal convergence result given in [45] to the new notation used here.

Theorem 2.3. *(the Huang theorem, [45]) Let $F : \Omega \subseteq X \longrightarrow Y$ be a twice continuously Fréchet differentiable operator defined on a nonempty open convex domain Ω of a Banach space X with values in a Banach space Y. Suppose that conditions (B1)-(B2)-(B3) are satisfied and $B(x_0, t^*) \subset \Omega$. Then, Newton's sequence, given by (1.1) and starting at x_0, converges to a solution x^* of $F(x) = 0$. Moreover, $x_n, x^* \in \overline{B(x_0, t^*)}$, for all $n \in \mathbb{N}$, and x^* is unique in $B(x_0, t^{**}) \cap \Omega$ if $t^* < t^{**}$ or in $\overline{B(x_0, t^*)}$ if $t^* = t^{**}$. Furthermore,*

$$\|x^* - x_n\| \le t^* - t_n, \quad n = 0, 1, 2, \dots,$$

where $\{t_n\}$ is defined in (2.3).

Proof. As we have indicated above, given the convergence of real sequence (2.3), it is sufficient to prove that the scalar function ϕ majorizes the operator F in the sense of Definition 1.26. The proof will be completed in three steps.

First, we prove the following property. If $x \in X$ satisfies $\|x - x_0\| \le t \le t^*$, then

$$\left\|[F'(x)]^{-1}\right\| \le -\frac{1}{\phi'(t)} \quad \text{and} \quad \|F''(x)\| \le \phi''(t). \tag{2.4}$$

In fact,

$$\|F''(x)\| \le \|F''(x_0)\| + \|F''(x) - F''(x_0)\| \le \delta + K\|x - x_0\| \le \delta + Kt = \phi''(t). \tag{2.5}$$

Since

$$I - \Gamma_0 F'(x) = -\Gamma_0 \left(F'(x) - F'(x_0) - F''(x_0)(x - x_0) + F''(x_0)(x - x_0)\right)$$
$$= -\Gamma_0 \int_0^1 \left(F''(x_0 + \tau(x - x_0)) - F''(x_0)\right) d\tau(x - x_0) - \Gamma_0 F''(x_0)(x - x_0)$$

and $\phi'(t) < 0$ in $[0, t^*]$, we obtain

$$\|I - \Gamma_0 F'(x)\| \le \beta \int_0^1 K\tau \, d\tau \|x - x_0\|^2 + \beta\delta\|x - x_0\| \le \frac{1}{2} K\beta t^2 + \beta\delta t = \beta\phi'(t) + 1 < 1,$$

so that there exists $[F'(x)]^{-1}$ and

$$\left\|[F'(x)]^{-1}\right\| \le \frac{\|\Gamma_0\|}{1 - \|I - \Gamma_0 F'(x)\|} \le -\frac{1}{\phi'(t)}, \tag{2.6}$$

by the Banach lemma on invertible operators.

Next, we prove that the function ϕ majorizes the operator F, in the sense of Definition 1.26; namely,

$$\|x_{n+1} - x_n\| \le t_{n+1} - t_n \quad \text{and} \quad x_n \in B(x_0, t^*), \tag{2.7}$$

for all $n = 0, 1, 2, \ldots$, where $\{t_n\}$ is defined in (2.3).

As $\|x_1 - x_0\| = \|\Gamma_0 F(x_0)\| = \eta = t_1 - t_0 < t^*$, then (2.7) holds for $n = 0$. If we now suppose that (2.7) is true for $n = 0, 1, 2, \ldots, j$, then

$$\|x_{j+1} - x_0\| \leq \sum_{i=1}^{j+1} \|x_i - x_{i-1}\| \leq \sum_{i=1}^{j+1} (t_i - t_{i-1}) = t_{j+1} - t_0 = t_{j+1} < t^*$$

and $\|x_j + \tau(x_{j+1} - x_j) - x_0\| \leq t_j + \tau(t_{j+1} - t_j) < t^*$. Since

$$\|F(x_{j+1})\| \leq \int_0^1 \|F''(x_j + \tau(x_{j+1} - x_j))\| (1 - \tau) \, d\tau \|x_{j+1} - x_j\|$$

$$\leq \int_0^1 \phi''(t_j + \tau(t_{j+1} - t_j))(1 - \tau) \, d\tau (t_{j+1} - t_j)^2$$

$$= \phi(t_{j+1}) \tag{2.8}$$

is derived from (2.4). Also because we can draw the conclusion that

$$\|\Gamma_{j+1}\| = \left\|[F'(x_{j+1})]^{-1}\right\| \leq -\frac{1}{\phi'(t_{j+1})},$$

by (2.6), then

$$\|x_{j+2} - x_{j+1}\| \leq \|\Gamma_{j+1}\| \|F(x_{j+1})\| \leq -\frac{\phi(t_{j+1})}{\phi'(t_{j+1})} = t_{j+2} - t_{j+1}.$$

In addition,

$$\|x_{j+2} - x_0\| \leq \|x_{j+2} - x_{j+1}\| + \|x_{j+1} - x_0\| \leq t_{j+2} - t_{j+1} + t_{j+1} - t_0 = t_{j+2} - t_0 < t^*.$$

Therefore, (2.7) holds for $n = j + 1$. The proof of (2.7) is completed by induction.

As a consequence, sequence (2.3) majorizes the sequence $\{x_n\}$ defined by Newton's method and, since $\lim_n t_n = t^*$, there exists $\lim_n x_n = x^*$ and $\|x^* - x_n\| \leq t^* - t_n$, for all $n = 0, 1, 2, \ldots$ The combination of this and (2.8) yields $F(x^*) = 0$.

Finally, we prove the uniqueness of solution x^*. Suppose that there exists a solution y^* of $F(x) = 0$, such that $y^* \neq x^*$, in $B(x_0, t^{**}) \cap \Omega$ if $t^* < t^{**}$ or in $\overline{B(x_0, t^*)}$ if $t^* = t^{**}$. First, we suppose that $t^* < t^{**}$. Then,

$$\|y^* - x_0\| = \rho(t^{**} - t_0) \quad \text{with} \quad \rho \in (0, 1).$$

We now suppose that $\|y^* - x_i\| \leq \rho^{2^i}(t^{**} - t_i)$, for $i = 0, 1, \ldots, n$, and use mathematical induction on n. So, from

$$y^* - x_{n+1} = y^* - x_n + \Gamma_n F(x_n)$$

$$= -\Gamma_n \left(F(y^*) - F(x_n) - F'(x_n)(y^* - x_n)\right)$$

$$= -\Gamma_n \int_{x_n}^{y^*} F''(z)(y^* - z) \, dz$$

$$= -\Gamma_n \int_0^1 F''(x_n + \tau(y^* - x_n))(y^* - x_n)^2(1 - \tau) \, d\tau$$

and taking into account $x = x_n + \tau(y^* - x_n)$ and $t = t_n + \tau(t^{**} - t_n)$, with $\tau \in [0,1]$, that
satisfy $\|x - x_0\| \leq t$, along with (2.5) and (2.6), it follows

$$\|y^* - x_{n+1}\| \leq \|\Gamma_n\| \left(\int_0^1 \phi''(t_n + \tau(t^{**} - t_n))(1 - \tau) \, d\tau \right) \|y^* - x_n\|^2$$

$$\leq -\frac{\mu}{\phi'(t_n)} \|y^* - x_n\|^2,$$

where $\mu = \int_0^1 \phi''(t_n + \tau(t^{**} - t_n))(1 - \tau) \, d\tau$.

On the other hand, we also have

$$t^{**} - t_{n+1} = -\frac{1}{\phi'(t_n)} \int_{t_n}^{t^{**}} \phi''(t)(t^{**} - t) \, dt = -\frac{\mu}{\phi'(t_n)}(t^{**} - t_n)^2.$$

Therefore,

$$\|y^* - x_{n+1}\| \leq \frac{t^{**} - t_{n+1}}{(t^{**} - t_n)^2} \|y^* - x_n\|^2 \leq \rho^{2^{n+1}}(t^{**} - t_{n+1})$$

and the induction is complete. As a consequence, $y^* = x^*$, since $\|y^* - x_n\| \to 0$ as $n \to +\infty$.

Second, if $t^* = t^{**}$ and y^* is another solution of $F(x) = 0$ in $\overline{B}(x_0, t^*)$, then $\|y^* - x_0\| \leq t^* - t_0$. Proceeding similarly to the previous case, we can prove by mathematical induction on n that $\|y^* - x_n\| \leq t^* - t_n$. Since $t^* = t^{**}$ and $\lim_n t_n = t^*$, the uniqueness of solution is now easy to follow. ∎

Remark 2.4. If we pay attention to the proof of the Huang theorem, we see that condition
(B2) is not necessary, since it is enough for the condition to be fulfilled

$$\|F''(x) - F''(x_0)\| \leq K_0 \|x - x_0\|, \quad \text{for} \quad x \in \Omega, \tag{2.9}$$

instead of condition (B2). Gutiérrez proves in [37] the semilocal convergence of Newton's
method under condition (2.9).

Remark 2.5. As Kantorovich does for real sequence (1.22), it is clear that Huang's polynomial,
(2.2), can be obtained by solving the problem of interpolation fitting given by conditions (B1)-
(B2). For this, we look for a cubic polynomial, since its second derivative is Lipschitz, and
fit its coefficients with conditions (B1)-(B2).

The Newton-Kantorovich theorem and the Huang theorem are not comparable with respect to the accessibility of solution, since the convergence of Newton's method can be established using the Newton-Kantorovich theorem or the Huang theorem indistinctly. In the
following example, we first see that the semilocal convergence of Newton's method is guaranteed from the Huang theorem, but not from the Newton-Kantorovich theorem. Next, the
semilocal convergence of Newton's method is guaranteed from both theorems, but the Huang
theorem gives better domains of existence and uniqueness of solution and better error bounds
than the Newton-Kantorovich theorem.

Example 2.6. Now, we apply the Huang theorem to Example 2.1. For (2.1), we have $\delta = 0$,
since $x_0(s) = 0$, and $K = \frac{3}{4}$. Then, $\phi(t) = \frac{t^3}{12} - t + \sin 1$ and, as a consequence, $t^* = 0.9479\ldots$
and $t^{**} = 2.2326\ldots$ are the two positive real zeros of $\phi(t)$. Thus, from the Huang theorem,

as conditions (B1)-(B3) are satisfied, we can approximate a solution $x^*(s)$. The domains of existence and uniqueness of solution are, respectively,

$$\{\nu \in \mathcal{C}([0,1]) : \|\nu(s) - x_0(s)\| \leq 0.9479\ldots\} \text{ and } \{\nu \in \mathcal{C}([0,1]) : \|\nu(s) - x_0(s)\| < 2.2326\ldots\}.$$

Next, we proceed as in Section 1.2 to construct the iterations of Newton's method. Then, after five iterations of Newton's method, we obtain the approximated solution $x^*(s) = (0.0737\ldots)s^2 + \sin s$ given in Table 2.1. Observe that we can deduce, from the sequence $\{\|[\mathcal{F}(x_n)](s)\|\}$ shown in Table 2.1, that $x^*(s)$ is a good approximation of a solution of (2.1). In Table 2.2, we show the errors $\|x^*(s) - x_n(s)\|$ and the error bounds. Moreover, see Figure 2.1, the approximated solution $x^*(s)$ lies within the existence domain of solution obtained above. Finally, remember that $x^*(s)$ is a solution that is beyond the scope of the Newton-Kantorovich theorem when the starting point $x_0(s) = 0$ is chosen.

n	$x_n(s)$	$\|[\mathcal{F}(x_n)](s)\|$
0	0	$8.4147\ldots \times 10^{-1}$
1	$\sin s$	$5.8927\ldots \times 10^{-2}$
2	$(0.07242800189221\ldots)s^2 + \sin s$	$1.0704\ldots \times 10^{-3}$
3	$(0.07379388480767\ldots)s^2 + \sin s$	$4.0016\ldots \times 10^{-7}$
4	$(0.07379439582054\ldots)s^2 + \sin s$	$5.6062\ldots \times 10^{-14}$
5	$(0.07379439582061\ldots)s^2 + \sin s$	

Table 2.1: Approximated solution $x^*(s)$ of (2.1) and $\{\|[\mathcal{F}(x_n)](s)\|\}$

n	$\|x^*(s) - x_n(s)\|$	$\|t^* - t_n\|$
0	$9.1526\ldots \times 10^{-1}$	$9.4795\ldots \times 10^{-1}$
1	$7.3794\ldots \times 10^{-2}$	$1.0647\ldots \times 10^{-1}$
2	$1.3663\ldots \times 10^{-3}$	$5.0765\ldots \times 10^{-3}$
3	$5.1101\ldots \times 10^{-7}$	$1.3693\ldots \times 10^{-5}$
4	$7.1591\ldots \times 10^{-14}$	$1.0054\ldots \times 10^{-10}$

Table 2.2: Absolute errors and error bounds

On the other hand, the solution $x^*(s)$ previously approximated by Newton's method can be also done, after four iterations, if the method starts at $x_0(s) = \sin s$. In this case, we can guarantee the semilocal convergence of Newton's method to $x^*(s)$ by the Newton-Kantorovich theorem and the Huang theorem, but the domains of existence and uniqueness of solution and the error bounds obtained by the Newton-Kantorovich theorem are improved by the Huang theorem as we can see in the following.

Taking into account $x_0(s) = \sin s$, the real positive zeros of Kantorovich's polynomial, $p(s) = 0.75s^2 - (0.7828\ldots)s + (0.0589\ldots)$, are $s^* = 0.0816\ldots$ and $s^{**} = 0.9622\ldots$ As a consequence, by the Newton-Kantorovich theorem, we obtain that the domains of existence and uniqueness of solutions are respectively

$$\{\nu \in B(0,2) : \|\nu(s) - x_0(s)\| \leq 0.0816\ldots\} \text{ and } \{\nu \in B(0,2) : \|\nu(s) - x_0(s)\| < 0.9622\ldots\}.$$

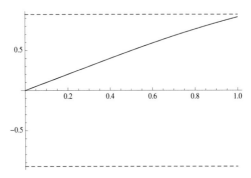

Figure 2.1: Graph (the solid line) of the approximated solution $x^*(s)$ of (2.1)

For the Huang theorem, we have that $\phi(t) = 0.125t^3 + (0.3155\ldots)t^2 - (0.7828\ldots)t + (0.0589\ldots)$, whose real positive zeros are $t^* = 0.0777\ldots$ and $t^{**} = 1.4834\ldots$, so that the domains of existence and uniqueness of solutions are respectively

$$\{\nu \in \mathcal{C}([0,1]) : \|\nu(s) - x_0(s)\| \leq 0.0777\ldots\} \text{ and } \{\nu \in \mathcal{C}([0,1]) : \|\nu(s) - x_0(s)\| < 1.4834\ldots\},$$

that improve those given by the Newton-Kantorovich theorem, since the domain of existence of solution is smaller (better location of the solution) and the domain of uniqueness of solution is bigger (better separation of solutions).

Finally, in Table 2.3, we can see that the error bounds $\{|t^* - t_n|\}$ obtained from the scalar sequence defined by Newton's method with ϕ (the Huang theorem) improves those, $\{|s^* - s_n|\}$, obtained from the scalar sequence defined by Newton's method with Kantorovich's polynomial p.

n	$\|x^*(s) - x_n(s)\|$	$\|t^* - t_n\|$	$\|s^* - s_n\|$
0	$7.3794\ldots \times 10^{-2}$	$7.7782\ldots \times 10^{-2}$	$8.1656\ldots \times 10^{-2}$
1	$1.3663\ldots \times 10^{-3}$	$2.5136\ldots \times 10^{-3}$	$6.3875\ldots \times 10^{-3}$
2	$5.1101\ldots \times 10^{-7}$	$2.9650\ldots \times 10^{-6}$	$4.5673\ldots \times 10^{-5}$
3	$7.1591\ldots \times 10^{-14}$	$4.1426\ldots \times 10^{-12}$	$2.3687\ldots \times 10^{-9}$

Table 2.3: Absolute errors and error bounds

2.1.2 Operators with Hölder second-derivative

A first natural extension of the semilocal convergence obtained in the Huang theorem to a more general situation is to assume, instead of condition (B2), that F'' satisfies a Hölder condition in Ω; i.e.:

$$\|F''(x) - F''(y)\| \leq K\|x - y\|^p, \quad p \in [0,1], \quad \text{for } x, y \in \Omega. \tag{2.10}$$

Observe that condition (2.10) is reduced to condition (B2) if $p = 1$.

We do not study this case because, below, we give a generalization that includes the Lipschitz and Hölder cases, conditions (B2) and (2.10), respectively.

2.1.3 Operators with ω-Lipschitz second-derivative

There are operators that do not satisfy any of the above conditions, as for example a nonlinear integral equation of Hammerstein-type (1.36) with $\mathcal{H}(x(t))$ defined in (1.39) and such that $\lambda_1 \neq 0$, $\lambda_2 \neq 0$, $p \in [0, 1]$ and $q \geq 3$; i.e.:

$$x(s) = u(s) + \int_a^b \mathcal{K}(s, t)\left(\lambda_1 \, x(t)^{2+p} + \lambda_2 \, x(t)^q\right) dt, \quad s \in [a, b]. \qquad (2.11)$$

Observe that solving equation (2.11) is equivalent to solving $\mathcal{F}(x) = 0$, where $\mathcal{F} : \mathcal{C}([a, b]) \longrightarrow \mathcal{C}([a, b])$ and

$$[\mathcal{F}(x)](s) = x(s) - u(s) - \int_a^b \mathcal{K}(s, t)\left(\lambda_1 \, x(t)^{2+p} + \lambda_2 \, x(t)^q\right) dt, \quad s \in [a, b].$$

Then,

$$\mathcal{F}'(x)y](s) = y(s) - \int_a^b \mathcal{K}(s, t)\left(\lambda_1(2 + p)x(t)^{1+p} + \lambda_2 q \, x(t)^{q-1}\right) y(t) \, dt,$$

$$[\mathcal{F}''(x)(yz)](s) = -\int_a^b \mathcal{K}(s, t)\left(\lambda_1(2 + p)(1 + p)x(t)^p + \lambda_2 q(q - 1)x(t)^{q-2}\right) z(t)y(t) \, dt.$$

As a consequence, if $q = 3$, it follows

$$\|\mathcal{F}''(x) - \mathcal{F}''(y)\| \leq S\left(|\lambda_1|(2 + p)(1 + p)\|x - y\|^p + 6|\lambda_2|\|x - y\|\right),$$

where $S = \left\|\int_a^b \mathcal{K}(s, t)\, dt\right\|$. Therefore, conditions (B2) and (2.10) are not easily satisfied and then we think of generalizing them.

By the last, we generalize the semilocal convergence results obtained previously to a still more general situation assuming, instead of condition (2.10), that F'' satisfies in Ω the following condition:

$$\|F''(x) - F''(y)\| \leq \omega(\|x - y\|), \quad \text{for all } x, y \in \Omega, \qquad (2.12)$$

where $\omega : [0, +\infty) \longrightarrow \mathbb{R}$ is a nondecreasing continuous function such that $\omega(0) = 0$.

Observe that condition (2.12) is reduced to conditions (B2) or (2.10) if $\omega(z) = Kz$ or $\omega(z) = Kz^p$, respectively.

We have seen in Section 1.1.2 that Kantorovich obtains majorizing sequence (1.22) from conditions (A1)-(A2) by solving a problem of interpolation fitting. We have just also seen that Huang does the same from conditions (B1)-(B2), see Remark 2.5. But, if we consider (B1) and (2.12), we cannot obtain a scalar function by interpolation fitting, as Kantorovich and Huang do, since (2.12) does not allow determining the class of functions where (B1) and (2.12) can be applied. To solve this problem, we proceed differently, without interpolation fitting, by solving an initial value problem and applying the modification of the majorant principle of Kantorovich seen in Section 1.1.4, as we indicate in Remark 1.31 to obtain the majorant function given by polynomial (1.23).

This new way of getting a majorant function has the advantage of being able to be generalized to condition (2.12).

First of all, our idea is to generalize conditions (B2) and (2.10). For this, the first thing we look for is to guarantee the existence of majorant functions for these situations. Next, taking into account what was said in Remarks 2.4 and 2.5 for the Huang study, we consider that condition (2.12) on the second derivative of the operator involved is centered at the starting point x_0 of Newton's method. Then, we consider that there exists a real function $f \in C^2([t_0, +\infty))$, with $t_0 \in \mathbb{R}$, which satisfies:

(D1) there exists $\Gamma_0 = [F'(x_0)]^{-1} \in \mathcal{L}(Y, X)$, for some $x_0 \in \Omega$, with $\|\Gamma_0\| \leq -\dfrac{1}{f'(t_0)}$

and $\|\Gamma_0 F(x_0)\| \leq -\dfrac{f(t_0)}{f'(t_0)}$; moreover, $\|F''(x_0)\| \leq f''(t_0)$,

(D2) $\|F''(x) - F''(x_0)\| \leq f''(t) - f''(t_0)$, for $\|x - x_0\| \leq t - t_0$, for $x \in \Omega$ and $t \in [t_0, +\infty)$.

Next, we use the modification of the majorant principle of Kantorovich given in Section 1.1.4 to prove the semilocal convergence of Newton's method under general conditions (D1)-(D2). For this, we look for a scalar function f that majorizes the operator F, in the sense of Definition 1.26, and construct a majorizing real sequence $\{t_n\}$ of Newton's sequence $\{x_n\}$ in the Banach space X. To obtain the sequence $\{t_n\}$ we use the previous real function $f(t)$ defined in $[t_0, +\infty) \subset \mathbb{R}$ as follows:

$$t_0 \text{ given,} \quad t_{n+1} = N_f(t_n) = t_n - \frac{f(t_n)}{f'(t_n)}, \quad n = 0, 1, 2, \ldots \tag{2.13}$$

We have seen in Section 1.1.2 that Kantorovich constructs a majorizing real sequence $\{s_n\}$ from the application of Newton's method to polynomial (1.23), so that this sequence converges to the smallest positive zero of the polynomial, see Theorem 1.21. Therefore, to construct a majorizing real sequence $\{t_n\}$ from $f(t)$, it is necessary that the function $f(t)$ has at least one zero t^*, such that $t^* \geq t_0$, and the sequence $\{t_n\}$ is nondecreasing and convergent to t^*. When this happens, the semilocal convergence of Newton's sequence $\{x_n\}$ is guaranteed in the Banach space X from the convergence of the scalar sequence $\{t_n\}$ defined in (2.13).

2.1.3.1 Existence of a majorizing sequence

To see the above-mentioned, we first study the conditions that the function f must satisfy to obtain a majorizing real sequence $\{t_n\}$ of type (2.13) that is convergent. If conditions (D1)-(D2) are satisfied and there exists a root $\alpha \in (t_0, +\infty)$ of $f'(t) = 0$ such that $f(\alpha) \leq 0$, then the equation $f(t) = 0$ has only one root t^* in (t_0, α). Indeed, if $f(\alpha) < 0$, as $f(t_0) > 0$, then $f(t)$ has at least one zero t^* in (t_0, α) by continuity. Besides, since $f''(t_0) \geq 0$, from (D2) it follows that $f''(t) \geq 0$ for $t \in (t_0, \alpha)$, so that $f'(t)$ is nondecreasing and $f'(t) < 0$ for $t \in (t_0, \alpha)$. Also, as $f'(t_0) < 0$, $f(t)$ is nonincreasing for $t \in [t_0, \alpha)$. As a consequence, t^* is the unique solution of $f(t) = 0$ in (t_0, α). On the other hand, if $f(\alpha) = 0$, then α is a double root of $f(t) = 0$ and we choose $t^* = \alpha$.

Taking into account that the function f has the same properties as Kantorovich's polynomial (1.23), since $f(t) > 0$, $f'(t) < 0$ and $f''(t) > 0$ in (t_0, α), the convergence of the real sequence $\{t_n\}$ is then guaranteed from the next theorem, whose proof is completely analogous to that of Theorem 1.21.

Theorem 2.7. *Suppose that there exist f such that conditions (D1)-(D2) are satisfied and there exists a root $\alpha \in (t_0, +\infty)$ of $f'(t) = 0$ such that $f(\alpha) \leq 0$. Then, the real sequence $\{t_n\}$, given in (2.13), is nondecreasing and converges to the unique solution t^* of $f(t) = 0$ in (t_0, α).*

From the previous notation, we consider the known degree of logarithmic convexity of a real function f (see [41]) as the function

$$L_f(t) = \frac{f(t)f''(t)}{f'(t)^2},$$

whose extension to Banach spaces can be used to construct majorizing real sequences for Newton's method in the Banach space X, as we can see below. In addition, we observe that $L_f(t) = N_f(t)$.

Now, we define the degree of logarithmic convexity in Banach spaces. For this, we suppose that $F : \Omega \subseteq X \longrightarrow Y$ is a twice continuously Fréchet differentiable operator in Ω and there exists the operator $[F'(x)]^{-1} : Y \longrightarrow X$. Moreover, since $F''(x) : X \times X \longrightarrow Y$, it follows $F''(x)[F'(x)]^{-1}F(x) \in \mathcal{L}(X,Y)$ and

$$L_F(x) : \Omega \xrightarrow{F''(x)[F'(x)]^{-1}F(x)} Y \xrightarrow{[F'(x)]^{-1}} \Omega.$$

In the same way that it is easy to see $L_f(t) = N_f'(t)$, we also have $L_F(x) = N_F'(x)$, as we can see in the following.

Lemma 2.8. *([55]) Let X and Y be Banach spaces. Consider two operators $Q : X \longrightarrow Y$ and $P : X \longrightarrow \mathcal{L}(Y,X)$, that are Fréchet differentiable at the point $x_0 \in X$, and, given $x \in X$, define the operator*

$$
\begin{array}{ccccc}
N_F(x) : & X & \xrightarrow{Q} & Y & \xrightarrow{P(x)} & X \\
& y & \longmapsto & Q(y) & \longmapsto & P(x)Q(y).
\end{array}
$$

Then, N_F is differentiable at x_0 and $N_F'(x_0) : X \longrightarrow X$ is a lineal operator given by

$$N_F'(x_0)[-] = \Big(P'(x_0)[-]Q\Big)(x_0) + P(x_0)Q'(x_0)[-].$$

From the previous lemma, we prove the following result.

Theorem 2.9. *If $N_F'(x_0)$ is defined as in Lemma 2.8, then $N_F'(x_0) = L_F(x_0)$.*

Proof. Let $G(x) = [\Gamma(x)F](x)$, where the operator $\Gamma : X \longrightarrow \mathcal{L}(X,Y)$ is such that $\Gamma(x) = [F'(x)]^{-1} \in \mathcal{L}(X,Y)$, for $x \in X$. In addition, $N_F'(x_0) = I - G'(x_0)$. Moreover, since

$$G'(x_0)(y) = [\Gamma'(x_0)(y)F](x_0) + [\Gamma(x_0)F'(x_0)](y) = [\Gamma'(x_0)(y)F](x_0) + y,$$

for $y \in X$, it follows that $N_F'(x_0)(z) = -[\Gamma'(x_0)(z)F](x_0)$.

On the other hand, we can write $\Gamma = TF'$, where

$$\Gamma : X \xrightarrow{F'} \mathcal{L}(X,Y) \xrightarrow{T} \mathcal{L}(Y,X),$$

so that $T : \mathcal{L}(X,Y) \longrightarrow \mathcal{L}(Y,X)$ is the inverse operator $T(F) = F^{-1}$. If we now apply the chain rule for differentiating compositions of operators, we have $\Gamma'(x_0) \in \mathcal{L}(X, \mathcal{L}(Y,X))$ and

$$\Gamma'(x_0)[-] = T'(F'(x_0))F''(x_0).$$

In addition, for $y \in X$,

$$\Gamma'(x_0)(y) = T'(F'(x_0))(F''(x_0)(y)) = -\Gamma(x_0)F''(x_0)(y)\Gamma(x_0) \in \mathcal{L}(Y,X).$$

Taking now into account the symmetry of the operator $F''(x_0)$, for $y \in X$, it follows that

$$\Gamma(x_0)F''(x_0)(y)\Gamma(x_0)F(x_0) = \Gamma(x_0)F''(x_0)\Gamma(x_0)F(x_0)(y) = L_F(x_0)(y).$$

The proof is complete. ∎

Now, it is easy to see that

$$x_{n+1} - x_n = \int_{x_{n-1}}^{x_n} N_F'(x)\, dx = \int_{x_{n-1}}^{x_n} L_F(x)\, dx,$$

where $\{x_n\}$ is Newton's sequence in the Banach space X, so that

$$\|x_{n+1} - x_n\| \le t_{n+1} - t_n, \quad n = 0, 1, 2, \ldots,$$

provided that $\|L_F(x)\| \le L_f(t)$ for $\|x - x_0\| \le t - t_0$.

The important and easy property given in Theorem 2.9 allows us to obtain a majorizing real sequence for Newton's method under conditions (D1)-(D2).

Note that the existence of the operator L_F in a domain seems a demanding condition, since it involves the existence of inverse operators $[F'(x)]^{-1}$ in any point x of the domain. However, the existence of the operator L_F is guaranteed under certain conditions for F''. In fact, Kantorovich's conditions for Newton's method are enough to guarantee the existence of the operator L_F, as we can see in the following result.

Theorem 2.10. *Let $F : \Omega \subseteq X \longrightarrow Y$ be a twice continuously Fréchet differentiable operator defined on a nonempty open convex domain Ω of a Banach space X with values in a Banach space Y. Suppose that the classic conditions of Kantorovich (A1)-(A2) are satisfied and $B\left(x_0, \frac{1}{M\beta}\right) \subset \Omega$ for all $x \in \Omega$. Then, the operator $L_F(x)$ exists for all $x \in B\left(x_0, \frac{1}{M\beta}\right) \subset \Omega$ and*

$$\|L_F(x)\| \le \frac{M\beta}{(1 - M\beta\|x - x_0\|)^2} \|\Gamma_0 F(x)\|. \tag{2.14}$$

Proof. From

$$\int_{x_0}^{x} \Gamma_0 F''(z)\, dz = \Gamma_0 F'(x) - I,$$

we have

$$\|\Gamma_0 F'(x) - I\| = \left\|\int_{x_0}^{x} \Gamma_0 F''(z)\, dz\right\| \le M\beta\|x - x_0\| < 1,$$

so that, from the Banach lemma on invertible operators, it follows that there exists $[\Gamma_0 F'(x)]^{-1}$, for all $x \in B\left(x_0, \frac{1}{M\beta}\right) \subset \Omega$, and

$$\|[\Gamma_0 F'(x)]^{-1}\| \le \frac{1}{1 - M\beta\|x - x_0\|}.$$

As a consequence, the operator $L_F(x)$ exists, since

$$L_F(x) = [\Gamma_0 F'(x)]^{-1}\Gamma_0 F''(x)[\Gamma_0 F'(x)]^{-1}\Gamma_0 F(x),$$

and (2.14) follows immediately. ∎

Before we see that the sequence $\{t_n\}$ given in (2.13) is a majorizing sequence of Newton's sequence $\{x_n\}$ given in the Banach space X, we give the following technical lemma that is used later.

Lemma 2.11. *Let $f \in C^2([t_0, +\infty))$ with $t_0 \in \mathbb{R}$. Suppose that conditions (D1)-(D2) are satisfied and there exits $\alpha \in (t_0, +\infty)$ such that $f'(\alpha) = 0$ and $B(x_0, \alpha - t_0) \subseteq \Omega$. Then, for $x \in B(x_0, \alpha - t_0)$, the operator $L_F(x) = \Gamma F''(x)\Gamma F(x)$ exists, where $\Gamma = [F'(x)]^{-1}$, and*

$$\|L_F(x)\| \leq \frac{f''(t)}{f'(t)^2}\|F(x)\| \quad for \quad \|x - x_0\| \leq t - t_0. \tag{2.15}$$

Proof. We start proving that the operator Γ exists for $x \in B(x_0, \alpha - t_0)$ and, for $t \in (t_0, \alpha)$ with $\|x - x_0\| \leq t - t_0$, we also have

$$\|\Gamma F'(x_0)\| \leq \frac{f'(t_0)}{f'(t)} \quad \text{and} \quad \|\Gamma\| \leq -\frac{1}{f'(t)}. \tag{2.16}$$

From

$$\|I - \Gamma_0 F'(x)\| = \left\|-\Gamma_0 \int_{x_0}^x F''(z)\,dz\right\|$$

$$\leq \|\Gamma_0\|\|F''(x_0)\|\|x - x_0\| + \|\Gamma_0\|\left\|\int_{x_0}^x (F''(z) - F''(x_0))\,dz\right\|$$

$$\leq -\frac{f''(t_0)}{f'(t_0)}(t - t_0) + \|\Gamma_0\|\left\|\int_0^1 (F''(x_0 + \tau(x - x_0)) - F''(x_0))(x - x_0)\,d\tau\right\|$$

and

$$\|z - x_0\| = \tau\|x - x_0\| \leq \tau(t - t_0) = t_0 + \tau(t - t_0) - t_0 = u - t_0,$$

where $z = x_0 + \tau(x - x_0)$ and $u = t_0 + \tau(t - t_0)$, with $\tau \in [0, 1]$, it follows from (D2) that

$$\|I - \Gamma_0 F'(x)\| \leq -\frac{f''(t_0)}{f'(t_0)}(t - t_0) - \frac{1}{f'(t_0)}\int_{t_0}^t (f''(u) - f''(t_0))\,du = 1 - \frac{f'(t)}{f'(t_0)} < 1,$$

since $f'(t)$ is nondecreasing and $f'(t_0) \leq f'(t) \leq 0$. As a consequence, by the Banach lemma on invertible operators, the operator Γ exists,

$$\|\Gamma\| \leq \frac{\|\Gamma_0\|}{1 - \|I - \Gamma_0 F'(x)\|} \leq -\frac{1}{f'(t)} \quad \text{and} \quad \|\Gamma F'(x_0)\| \leq \frac{1}{1 - \|I - \Gamma_0 F'(x)\|} \leq \frac{f'(t_0)}{f'(t)}.$$

Therefore (2.16) holds.

On the other hand, if $x \in B(x_0, \alpha - t_0)$ and $t \in [t_0, \alpha)$ are such that $\|x - x_0\| \leq t - t_0$, we

$$\|F''(x)\| \leq \|F''(x_0)\| + \|F''(x) - F''(x_0)\| \leq f''(t_0) + f''(t) - f''(t_0) = f''(t) \tag{2.17}$$

and (2.15) holds.   ∎

Next, from the following lemma, we see that the sequence $\{t_n\}$ given in (2.13) is a majorizing sequence of Newton's sequence $\{x_n\}$ given in the Banach space X.

Lemma 2.12. *Under the hypotheses of Lemma 2.11, the following items are true for all $n \geq 0$:*

(i_n) $x_n \in B(x_0, t^* - t_0) \subset \Omega$,

(ii_n) $\|\Gamma_0 F(x_n)\| \leq -\dfrac{f(t_n)}{f'(t_0)}$,

(iii_n) $\|x_{n+1} - x_n\| \leq t_{n+1} - t_n$.

Proof. We prove (i_n)-(ii_n)-(iii_n) by mathematical induction on n. First, x_0 given, it is clear that x_1 is well-defined and

$$\|x_1 - x_0\| = \|\Gamma_0 F(x_0)\| \leq -\frac{f(t_0)}{f'(t_0)} = t_1 - t_0 < t^* - t_0.$$

Thus, (i_0)-(ii_0)-(iii_0) hold. Next, we suppose that (i_j)-(ii_j)-(iii_j) are true for $j = 0, 1, \ldots, n-1$ and prove (i_n)-(ii_n)-(iii_n).

As $x_n = x_{n-1} - [F'(x_{n-1})]^{-1} F(x_{n-1})$, it is clear that x_n is well-defined, since $[F'(x_{n-1})]^{-1}$ exists by the last lemma. Moreover,

$$\|x_n - x_0\| \leq \|x_n - x_{n-1}\| + \|x_{n-1} - x_{n-2}\| + \cdots + \|x_1 - x_0\|$$
$$\leq t_n - t_{n-1} + t_{n-1} - t_{n-2} + \cdots + t_1 - t_0 = t_n - t_0 < t^* - t_0,$$

so that $x_n \in B(x_0, t^* - t_0) \subset \Omega$ and (i_n) holds.

After that, we consider $x = x_{n-1} + s(x_n - x_{n-1})$, with $s \in [0, 1]$, so that $x - x_{n-1} = s(x_n - x_{n-1}) = -s[F'(x_{n-1})]^{-1} F(x_{n-1})$ and $\|x - x_{n-1}\| = s\|x_n - x_{n-1}\| \leq s(t_n - t_{n-1})$. Then, taking into account Taylor's series, we write

$$\|\Gamma_0 F(x)\| = \left\| \Gamma_0 F(x_{n-1}) + \Gamma_0 F'(x_{n-1})(x - x_{n-1}) + \int_{x_{n-1}}^{x} \Gamma_0 F''(z)(x - z)\, dz \right\|$$

$$\leq (1-s)\|\Gamma_0 F(x_{n-1})\| + \frac{1}{2}\|\Gamma_0\|\|F''(x_0)\|\|x - x_{n-1}\|^2$$
$$+ \|\Gamma_0\| \int_{x_{n-1}}^{x} \|F''(z) - F''(x_0)\| \|x - z\|\, dz$$

$$= (1-s)\|\Gamma_0 F(x_{n-1})\| + \frac{1}{2}\|\Gamma_0\|\|F''(x_0)\|\|x - x_{n-1}\|^2$$
$$+ \|\Gamma_0\| \int_0^1 \|F''(x_{n-1} + \tau(x - x_{n-1})) - F''(x_0)\| \|x - x_{n-1}\|^2 (1 - \tau)\, d\tau.$$

Moreover, as $\|x - x_{n-1}\| \leq s(t_n - t_{n-1}) \leq t - t_{n-1}$, for $t = t_{n-1} + s(t_n - t_{n-1})$, and $z = x_{n-1} + \tau(x - x_{n-1})$ with $\tau \in [0, 1]$, we have

$$\|z - x_0\| \leq \|x_{n-1} - x_0\| + \tau\|x - x_{n-1}\| \leq t_{n-1} - t_0 + \tau(t - t_{n-1}) = u - t_0,$$

where $u = t_{n-1} + \tau(t - t_{n-1})$, and

$$\|\Gamma_0 F(x)\| \leq -(1-s)\frac{f(t_{n-1})}{f'(t_0)} - \frac{1}{2}\frac{f''(t_0)}{f'(t_0)}(t - t_{n-1})^2 -$$

$$- \frac{1}{f'(t_0)} \int_0^1 (f''(t_{n-1} + \tau(t - t_{n-1})) - f''(t_0))(t_n - t_{n-1})^2 (1 - \tau)\, d\tau$$

$$= -\frac{1}{f'(t_0)} \left(f(t_{n-1}) + f'(t_{n-1})(t - t_{n-1}) + \int_{t_{n-1}}^{t} f''(u)(t - u)\, du \right)$$

$$= -\frac{f(t)}{f'(t_0)}.$$

If we now choose $s = 1$ above, we obtain $x = x_n$, $t = t_n$ and $\|\Gamma_0 F(x_n)\| \leq -\dfrac{f(t_n)}{f'(t_0)}$, so that (ii_n) holds.

To prove (iii_n), we first see that $\|L_F(x)\| \leq \dfrac{f(t)f''(t)}{f'(t)^2} = L_f(t)$. First, from Lemma 2.11, we have that the operators Γ and $L_F(x)$ exist. Second, if we now write $L_F(x) = \Gamma F''(x)\Gamma F'(x_0)\Gamma_0 F(x)$, it follows, from (2.16) and (2.17), that

$$\|L_F(x)\| \leq -\frac{f''(t)}{f'(t)}\frac{f'(t_0)}{f'(t)}\|\Gamma_0 F(x)\|,$$

with $x = x_{n-1} + s(x_n - x_{n-1})$, $t = t_{n-1} + s(t_n - t_{n-1})$ and $s \in [0,1]$, since $\|x - x_{n-1}\| = s\|x_n - x_{n-1}\| \leq s(t_n - t_{n-1})$ and

$$\|x - x_0\| \leq \|x_{n-1} - x_0\| + s\|x_n - x_{n-1}\| \leq t_{n-1} - t_0 + s(t_n - t_{n-1})$$
$$= t_{n-1} + s(t_n - t_{n-1}) - t_0 = t - t_0.$$

In addition, $\|L_F(x)\| \leq L_f(t)$. Finally,

$$\|x_{n+1} - x_n\| = \left\|\int_{x_{n-1}}^{x_n} L_F(x)\, dx\right\|$$
$$\leq \int_0^1 \|L_F(x_{n-1} + \tau(x_n - x_{n-1}))\|\, \|x_n - x_{n-1}\|\, d\tau$$
$$\leq \int_0^1 L_f(t_{n-1} + \tau(t_n - t_{n-1}))\, (t_n - t_{n-1})\, d\tau$$
$$= \int_{t_{n-1}}^{t_n} L_f(u)\, du$$
$$= t_{n+1} - t_n,$$

since $x = x_{n-1} + \tau(x_n - x_{n-1})$ with $\tau \in [0,1]$ and

$$\|x - x_0\| \leq \|x_{n-1} - x_0\| + \tau\|x_n - x_{n-1}\| \leq t_{n-1} + \tau(t_n - t_{n-1}) - t_0 = t - t_0,$$

where $t = t_{n-1} + \tau(t_n - t_{n-1})$. ∎

Once we have seen that (2.13) is a majorizing sequence of (1.1), we are ready to prove the semilocal convergence of (1.1) in the Banach space X.

Theorem 2.13. (General semilocal convergence, [26]) *Let $F : \Omega \subseteq X \longrightarrow Y$ be a twice continuously Fréchet differentiable operator defined on a nonempty open convex domain Ω of a Banach space X with values in a Banach space Y. Suppose that there exist $f \in \mathcal{C}^2([t_0, +\infty))$, with $t_0 \in \mathbb{R}$ and such that (D1)-(D2) are satisfied, and a root $\alpha \in (t_0, +\infty)$ of $f'(t) = 0$ such that $f(\alpha) \leq 0$, and $B(x_0, t^* - t_0) \subset \Omega$. Then, Newton's sequence, given by (1.1) and starting at x_0, converges to a solution x^* of $F(x) = 0$. Moreover, $x_n, x^* \in \overline{B}(x_0, t^* - t_0)$, for all $n \in \mathbb{N}$, and*

$$\|x^* - x_n\| \leq t^* - t_n, \quad n = 0, 1, 2, \ldots,$$

where $\{t_n\}$ is defined in (2.13).

Proof. Observe that $\{x_n\}$ is convergent, since $\{t_n\}$ is a majorizing real sequence of $\{x_n\}$ and convergent. Moreover, as $\lim_n t_n = t^*$, if $x^* = \lim_n x_n$, then $\|x^* - x_n\| \leq t^* - t_n$, for all $n = 0, 1, 2, \ldots$ Furthermore, from item (ii_n) of the last lemma, we have $\|\Gamma_0 F(x_n)\| \leq -\frac{f(t_n)}{f'(t_0)}$, for all $n = 0, 1, 2, \ldots$ Then, by letting $n \to +\infty$ in the last inequality, it follows $F(x^*) = 0$ by the continuity of F. ∎

2.1.3.2 Existence of a majorant function

Until now, we have supposed that there exists $f \in C^2([t_0, +\infty))$, with $t_0 \in \mathbb{R}$, such that (D1)-(D2) are satisfied. Now, we see that this function exists and we can then give a semilocal convergence result for Newton's method. For this, we suppose the following conditions:

(U1) There exists $\Gamma_0 = [F'(x_0)]^{-1} \in \mathcal{L}(Y, X)$, for some $x_0 \in \Omega$, with $\|\Gamma_0\| \leq \beta$ and $\|\Gamma_0 F(x_0)\| \leq \eta$; moreover, $\|F''(x_0)\| \leq \delta$.

(U2) There exists a continuous and nondecreasing function $\omega_0 : [0, +\infty) \longrightarrow \mathbb{R}$ such that $\|F''(x) - F''(x_0)\| \leq \omega_0(\|x - x_0\|)$, for $x \in \Omega$, and $\omega_0(0) = 0$.

After that, we look for a scalar function that majorizes the operator F in the sense of Definition 1.26. As we have seen above, this function depends on the starting point x_0, so that, to apply Theorem 2.13, we consider

$$\|F''(x) - F''(x_0)\| \leq \omega_0(\|x - x_0\|) \leq \omega_0(t - t_0) = f''(t) - f''(t_0) \quad \text{for} \quad \|x - x_0\| \leq t - t_0$$

and $f''(t) = \omega_0(t - t_0) + f''(t_0)$. If $f''(t_0) = \delta$, we can find a real function f that majorizes the operator F as the unique solution of the initial value problem

$$\begin{cases} y''(t) = \delta + \omega_0(t - t_0), \\ y(t_0) = \dfrac{\eta}{\beta}, \quad y'(t_0) = -\dfrac{1}{\beta}, \end{cases}$$

since, from (D1) and (U1), we can choose $-\dfrac{1}{f'(t_0)} = \beta$ and $-\dfrac{f(t_0)}{f'(t_0)} = \dfrac{\eta}{\beta}$. In addition, we can establish the following result.

Theorem 2.14. *Suppose that the function $\omega_0(t - t_0)$ is continuous for all $t \in [t_0, +\infty)$. Then, for any nonnegative real numbers $\beta \neq 0$, η and δ, there exists only one solution $\varphi(t)$ of the last initial value problem in $[t_0, +\infty)$; that is:*

$$\varphi(t) = \int_{t_0}^{t} \int_{t_0}^{\theta} \omega_0(\xi - t_0) \, d\xi \, d\theta + \frac{\delta}{2}(t - t_0)^2 - \frac{t - t_0}{\beta} + \frac{\eta}{\beta}. \tag{2.18}$$

To apply Theorem 2.13, the equation $\varphi(t) = 0$ must have at least one root greater than t_0, so that we have to guarantee the convergence of the real sequence $\{t_n\}$ defined in (2.13) with $f(t) \equiv \varphi(t)$ to this root. We then study the function $\varphi(t)$ defined in (2.18).

Theorem 2.15. *Let φ and ω_0 be the functions defined respectively in (2.18) and (U2).*

(a) *There exists only one positive solution $\alpha > t_0$ of the equation*

$$\varphi'(t) = \int_{t_0}^{t} \omega_0(\xi - t_0) \, d\xi + \delta(t - t_0) - \frac{1}{\beta} = 0, \tag{2.19}$$

which is the unique minimum of $\varphi(t)$ in $[t_0, +\infty)$ and $\varphi(t)$ is nonincreasing in $[t_0, \alpha)$.

(b) *If $\varphi(\alpha) \leq 0$, then the equation $\varphi(t) = 0$ has at least one root in $[t_0, +\infty)$. Moreover, if t^* is the smallest root of $\varphi(t) = 0$ in $[t_0, +\infty)$, we have $t_0 < t^* \leq \alpha$.*

Proof. First, as $\varphi'(\alpha) = \int_{t_0}^{\alpha} \omega_0(\xi - t_0)\, d\xi + \delta(\alpha - t_0) - \dfrac{1}{\beta} = 0$ and $\varphi''(t) = \omega_0(t - t_0) + \delta \geq 0$ in $[t_0 + \infty)$, then α is a minimum of φ in $[t_0, +\infty)$. Moreover, as φ is convex in $[t_0, +\infty)$, then α is the unique minimum of φ in $[t_0, +\infty)$ and φ' is nondecreasing in $[t_0, +\infty)$.

Second, as $\varphi'(t_0) = -\dfrac{1}{\beta} < 0$, $\varphi'(\alpha) = 0$ and φ' is nondecreasing in $[0, +\infty)$, then $\varphi'(t) < 0$ in $[t_0, \alpha)$, so that φ is nonincreasing in $[t_0, \alpha)$.

Third, if $\varphi(\alpha) < 0$, as $\varphi(t_0) = \dfrac{\eta}{\beta} > 0$ and φ is continuous, then φ has at least one zero t^* in (t_0, α). As φ is nonincreasing in $[t_0, \alpha)$, then t^* is the unique zero of φ in (t_0, α).

Finally, if $\varphi(\alpha) = 0$, then α is a double zero of φ and $t^* = \alpha$. ∎

2.1.3.3 Semilocal convergence

Condition (b) of Theorem 2.15, along with conditions (U1)-(U2), allow proving the semilocal convergence of Newton's method, since the sequence, $\{t_{n+1} = N_\varphi(t_n)\}$, for any $t_0 \in [0, \alpha)$, is convergent by Theorem 1.21. In addition, we consider:

(U3) $\varphi(\alpha) \leq 0$, where φ is the function given in (2.18) and α is the unique positive solution of $\varphi'(t) = 0$ such that $\alpha > t_0$, and $B(x_0, t^* - t_0) \subset \Omega$, where t^* is the smallest positive solution of $\varphi(t) = 0$ in $[t_0, +\infty)$.

Taking into account the hypotheses of Theorem 2.15, function (2.18) satisfies the conditions of Theorem 2.13 and the semilocal convergence of Newton's method is then guaranteed in the Banach space X if conditions (U1)-(U2)-(U3) are satisfied. In particular, we have the following theorem, whose proof follows immediately from Theorem 2.13.

Theorem 2.16. *Let $F : \Omega \subseteq X \longrightarrow Y$ be a twice continuously Fréchet differentiable operator defined on a nonempty open convex domain Ω of a Banach space X with values in a Banach space Y and $\varphi(t)$ be the function defined in (2.18). Suppose that (U1)-(U2)-(U3) are satisfied. Then, Newton's sequence $\{x_n\}$, given by (1.1), converges to a solution x^* of $F(x) = 0$ starting at x_0. Moreover, $x_n, x^* \in \overline{B(x_0, t^*)}$, for all $n \in \mathbb{N}$, and*

$$\|x^* - x_n\| \leq t^* - t_n, \quad n = 0, 1, 2, \ldots,$$

where $t_n = t_{n-1} - \dfrac{\varphi(t_{n-1})}{\varphi'(t_{n-1})}$, with $n \in \mathbb{N}$, t_0 given and $\varphi(t)$ defined in (2.18).

2.1.3.4 Uniqueness of solution and order of convergence

After proving the semilocal convergence of Newton's method and locating the solution x^*, we prove the uniqueness of x^*. Note that if $f(t)$ has two real zeros t^* and t^{**} such that $t_0 < t^* \leq t^{**}$, then the uniqueness of solution follows from the next theorem.

Theorem 2.17. *Under the hypotheses of Theorem 2.16, the solution x^* is unique in $B(x_0, t^{**} - t_0) \cap \Omega$ if $t^* < t^{**}$ or in $\overline{B(x_0, t^* - t_0)}$ if $t^* = t^{**}$.*

Proof. Suppose that there exists a solution y^* of $F(x) = 0$, such that $y^* \neq x^*$, in $B(x_0, t^{**} - t_0) \cap \Omega$ if $t^* < t^{**}$ or in $\overline{B(x_0, t^* - t_0)}$ if $t^* = t^{**}$. Then,

$$\|y^* - x_0\| = \rho(t^{**} - t_0) \text{ with } \rho \in (0, 1) \text{ if } x^* \in B(x_0, t^{**}) \cap \Omega,$$
$$\|y^* - x_0\| \leq t^* - t_0 \text{ if } x^* \in \overline{B(x_0, t^*)}.$$

In addition, from

$$
\begin{aligned}
y^* - x_{n+1} &= y^* - x_n + \Gamma_n F(x_n) \\
&= -\Gamma_n \left(F(y^*) - F(x_n) - F'(x_n)(y^* - x_n) \right) \\
&= -\Gamma_n \int_{x_n}^{y^*} F''(z)(y^* - z)\, dz \\
&= -\Gamma_n \int_0^1 F''(x_n + \tau(y^* - x_n))(y^* - x_n)^2 (1 - \tau)\, d\tau \\
&= -\Gamma_n \left(\int_0^1 \left(F''(x_n + \tau(y^* - x_n)) - F''(x_0) \right) (y^* - x_n)^2 (1 - \tau)\, d\tau \right. \\
&\qquad \left. + \frac{1}{2} F''(x_0)(y^* - x_n)^2 \right),
\end{aligned}
$$

taking into account (U2) and $\omega_0(\|x - x_0\|) \le \omega_0(t - t_0) = \varphi''(t) - \varphi''(t_0)$, for $\|x - x_0\| \le t - t_0$, and following a similar procedure to that given in the Huang theorem, Theorem 2.3, for uniqueness of solution, where mathematical induction on n is invoked, we obtain

$$
\|y^* - x_j\| \le \rho^{2^j}(t^{**} - t_j) \text{ if } x^* \in B(x_0, t^{**}) \cap \Omega,
$$
$$
\|y^* - x_j\| \le t^* - t_j \text{ if } x^* \in \overline{B(x_0, t^*)},
$$

for all positive integer j. Therefore, $\|y^* - x_j\| \to 0$ as $j \to +\infty$, which yields $y^* = x^*$. ∎

Note that the uniqueness of solution established in Theorem 2.17 includes the uniqueness of solution given by Huang (the Huang theorem) when $\varphi(t)$ is reduced to $\phi(t)$ and $t_0 = 0$.

We finish this section by seeing the quadratic convergence of Newton's method under conditions (U1)-(U2)-(U3). We obtain the following theorem from Ostrsowski's technique [65].

Notice first that if $\varphi(t)$ has two real zeros t^* and t^{**} such that $t_0 < t^* \le t^{**}$, we can then write

$$
\varphi(t) = (t^* - t)(t^{**} - t)g(t)
$$

with $g(t^*) \ne 0$ and $g(t^{**}) \ne 0$.

Theorem 2.18. *Under the hypotheses of Theorem 2.16 and assuming that $\varphi(t) = (t^* - t)(t^{**} - t)g(t)$, where $g(t)$ is a function such that $g(t^*) \ne 0$ and $g(t^{**}) \ne 0$, we have the following:*

(a) *If $t^* < t^{**}$ and there exist $m_1 > 0$ and $M_1 > 0$ such that $m_1 \le \min\{Q_1(t); t \in [t_0, t^*]\}$ and $M_1 \ge \max\{Q_1(t); t \in [t_0, t^*]\}$, where $Q_1(t) = \frac{(t^{**}-t)g'(t)-g(t)}{(t^*-t)g'(t)-g(t)}$, then*

$$
\frac{(t^{**} - t^*)\theta^{2^n}}{m_1 - \theta^{2^n}} \le t^* - t_n \le \frac{(t^{**} - t^*)\Delta^{2^n}}{M_1 - \Delta^{2^n}}, \quad n \ge 0,
$$

where $\theta = \frac{t^*}{t^{**}} m_1$, $\Delta = \frac{t^*}{t^{**}} M_1$ and provided that $\theta < 1$ and $\Delta < 1$.

(b) *If $t^* = t^{**}$ and there exist $m_2 > 0$ and $M_2 > 0$ such that $m_2 \le \min\{Q_2(t); t \in [t_0, t^*]\}$ and $M_2 \ge \max\{Q_2(t); t \in [t_0, t^*]\}$, where $Q_2(t) = \frac{(t^*-t)g'(t)-g(t)}{(t^*-t)g'(t)-2g(t)}$, then*

$$
m_2^n t^* \le t^* - t_n \le M_2^n t^*, \quad n \ge 0,
$$

provided that $m_2 < 1$ and $M_2 < 1$.

Proof. Let $t^* < t^{**}$ and denote $a_n = t^* - t_n$ and $b_n = t^{**} - t_n$ for all $n \geq 0$ Then

$$\varphi(t_n) = a_n b_n g(t_n), \quad \varphi'(t_n) = a_n b_n g'(t_n) - (a_n + b_n)g(t_n)$$

and

$$a_{n+1} = t^* - t_{n+1} = t^* - t_n + \frac{\varphi(t_n)}{\varphi'(t_n)} = \frac{a_n^2 (b_n g'(t_n) - g(t_n))}{a_n b_n g'(t_n) - (a_n + b_n)g(t_n)}.$$

From $\dfrac{a_{n+1}}{b_{n+1}} = \dfrac{a_n^2 (b_n g'(t_n) - g(t_n))}{b_n^2 (a_n g'(t_n) - g(t_n))}$, it follows

$$m_1 \left(\frac{a_n}{b_n}\right)^2 \leq \frac{a_{n+1}}{b_{n+1}} \leq M_1 \left(\frac{a_n}{b_n}\right)^2.$$

In addition,

$$\frac{a_{n+1}}{b_{n+1}} \leq M_1^{2^{n+1}-1} \left(\frac{a_0}{b_0}\right)^{2^{n+1}} = \frac{\Delta^{2^{n+1}}}{M_1} \quad \text{and} \quad \frac{a_{n+1}}{b_{n+1}} \geq m_1^{2^{n+1}-1} \left(\frac{a_0}{b_0}\right)^{2^{n+1}} = \frac{\theta^{2^{n+1}}}{m_1}.$$

Taking then into account that $b_{n+1} = (t^{**} - t^*) + a_{n+1}$, it follows:

$$\frac{(t^{**} - t^*)\theta^{2^{n+1}}}{m_1 - \theta^{2^{n+1}}} \leq t^* - t_{n+1} \leq \frac{(t^{**} - t^*)\Delta^{2^{n+1}}}{M_1 - \Delta^{2^{n+1}}}.$$

If $t^* = t^{**}$, then $a_n = b_n$ and

$$a_{n+1} = \frac{a_n (a_n g'(t) - g(t))}{a_n g'(t) - 2g(t_n)}.$$

As a consequence, $m_2 a_n \leq a_{n+1} \leq M_2 a_n$ and

$$m_2^{n+1} t^* \leq t^* - t_{n+1} \leq M_2^{n+1} t^*.$$

The proof is complete. ∎

From the last theorem, it follows that the convergence of Newton's method, under conditions (B1)-(B2), is quadratic if $t^* < t^{**}$ and linear if $t^* = t^{**}$.

Remark 2.19. Provided that the majorant function involved in the semilocal convergence of Newton's method can be factored as $(t^* - t)(t^{**} - t)g(t)$ with $g(t^*) \neq 0$ and $g(t^{**}) \neq 0$, we can obtain a result about the order of convergence of this method as that given in Theorem 2.18. In addition, from the proof of this theorem, we deduce that the order of convergence of the method is independent of the majorant function. For this reason, from now on, we omit the proofs of the order of convergence of Newton's method appearing in the textbook.

2.1.3.5 Applications

As we have indicated previously, the main aim of Section 2.1 is to require different conditions to the operator F'' than condition (A2) of the Newton-Kantorovich theorem, so that the domain of starting points for Newton's method is modified with respect to that of the Newton-Kantorovich theorem. In the following, we see three different situations where Kantorovich's study is improved, for one reason or another, from conditions (U1)-(U2)-(U3) and

Theorem 2.16. The three situations are presented in two examples, where two particular nonlinear integral equations of Hammerstein-type (2.11), with $q = 3$, are considered. In addition, in the three situations, the study of Huang is not applicable, since we cannot use the Huang theorem to guarantee the convergence of Newton's method as a result of the operator F'' is not Lipschitz continuous in the domain.

In the first example, we cannot apply the Newton-Kantorovich theorem because solutions of the integral equation given cannot be located previously. In the second example, two situations are taken into account. First, a solution of the integral equation given can be located previously, so that the Newton-Kantorovich theorem can be applied, but Theorem 2.16 gives better domains of existence and uniqueness of the solution, as well as better error bounds, than the Newton-Kantorovich theorem. And second, we apply Theorem 2.16 to guarantee the convergence of Newton's method to another solution of the integral equation that the Newton-Kantorovich theorem cannot establish, since we cannot locate the solution previously in an unbounded region.

In practice, we usually choose $t_0 = 0$, since the function φ has the same property of translation as Kantorovich's polynomial (see Remark 1.18) and majorizes, independent of the value of t_0, the operator F in the sense of Definition 1.26.

Example 2.20. As we have written in Remark 1.32, to solve the difficulty of applying condition (A2) of Kantorovich, a common alternative is to locate the solutions in a domain Ω and look for a bound for $\|F''(x)\|$ in Ω. In the next example we see that we cannot use this alternative either, because we cannot find, a priori, a domain Ω which contains solutions of the equation.

We consider a nonlinear integral equation of Hammerstein-type defined in (1.36)-(1.39) with $a = 0$, $b = 1$, $u(s) = s$, $\mathcal{K}(s,t) = e^{-(s+t)}$ and $\mathcal{H}(x(t)) = \frac{1}{3}x(t)^{11/5} + \frac{1}{4}x(t)^3$; i.e.:

$$x(s) = s + \int_0^1 e^{-(s+t)}\left(\frac{1}{3}x(t)^{11/5} + \frac{1}{4}x(t)^3\right) dt, \quad s \in [0,1]. \tag{2.20}$$

Solving equation (2.20) is equivalent to solving $\mathcal{F}(x) = 0$, where $\mathcal{F} : \mathcal{C}([0,1]) \longrightarrow \mathcal{C}([0,1])$ and

$$[\mathcal{F}(x)](s) = x(s) - s - \int_0^1 e^{-(s+t)}\left(\frac{1}{3}x(t)^{11/5} + \frac{1}{4}x(t)^3\right) dt.$$

Moreover,

$$[\mathcal{F}'(x)y](s) = y(s) - \int_0^1 e^{-(s+t)}\left(\frac{11}{15}x(t)^{6/5} + \frac{3}{4}x(t)^2\right) y(t)\, dt,$$

$$[\mathcal{F}''(x)(yz)](s) = \int_0^1 e^{-(s+t)}\left(\frac{22}{25}x(t)^{1/5} + \frac{3}{2}x(t)\right) z(t)y(t)\, dt$$

and

$$\|\mathcal{F}''(x) - \mathcal{F}''(y)\| \le \frac{e-1}{50\,e}\left(44\|x-y\|^{1/5} + 75\|x-y\|\right) = \omega(\|x-y\|).$$

Observe that the solutions $x^*(s)$ of the equation (2.20) must satisfy

$$\|x^*(s)\| - 1 - \frac{e-1}{e}\left(\frac{1}{3}\|x^*(s)\|^{11/5} + \frac{1}{4}\|x^*(s)\|^3\right) \le 0,$$

which does not imply any restriction on $\|x^*(s)\|$, so that we cannot locate a domain Ω where $\|\mathcal{F}''(x)\|$ is bounded and contains $x^*(s)$. As a consequence, we cannot guarantee the semilocal convergence of Newton's method to a solution of (2.20) from the Newton-Kantorovich theorem.

However, we can guarantee the semilocal convergence of Newton's method from Theorem 2.16. Indeed, if we choose $x_0(s) = s$, then $\beta = 1.3461\ldots$, $\eta = 0.1050\ldots$, $\delta = 1.5044\ldots$, $\omega_0(t) = \omega(t) = \frac{e-1}{50\,e}(44\sqrt[5]{t} + 75t)$ and

$$\varphi(t) = (0.0780\ldots) - (0.7428\ldots)t + (0.7522\ldots)t^2 + (0.2107\ldots)t^{11/5} + (0.1580\ldots)t^3$$

with $t_0 = 0$. This function $\varphi(t)$ has two positive real zeros $t^* = 0.1238\ldots$ and $t^{**} = 0.5899\ldots$ and the hypotheses of Theorem 2.16 are satisfied, since $\varphi(\alpha) = -0.0622\ldots \leq 0$, where $\alpha = 0.3616\ldots$ is the unique positive solution of $\varphi'(t) = 0$, and $B(s, 0.1238\ldots) \subset C([0,1])$. As a consequence, we can approximate a solution $x^*(s)$ by Newton's method. The domains of existence and uniqueness of solution are, respectively

$$\{\nu \in C([0,1]) : \|\nu(s) - x_0(s)\| \leq 0.1238\ldots\} \text{ and } \{\nu \in C([0,1]) : \|\nu(s) - x_0(s)\| < 0.5899\ldots\}.$$

Next, we proceed as in Section 1.2 to construct the iterations of Newton's method. Then, after four iterations of Newton's method, we obtain the approximated solution $x^*(s) = (0.0941\ldots)\,e^{-s} + s$ given in Table 2.4. Observe, from the sequence $\{\|[\mathcal{F}(x_n)](s)\|\}$, which is also given in Table 2.4, that $x^*(s)$ is a good approximation of a solution of (2.20). In Table 2.5, we show the errors $\|x^*(s) - x_n(s)\|$ and the error bounds. In Table 2.6, we show the sequences $\left\{\frac{(t^{**}-t^*)\theta^{2^n}}{m_1 - \theta^{2^n}}\right\}$ and $\left\{\frac{(t^{**}-t^*)\Delta^{2^n}}{M_1 - \Delta^{2^n}}\right\}$ of Theorem 2.18, which are denoted respectively by $\{d_n\}$ and $\{e_n\}$, and see that the a priori error bounds obtained from this theorem satisfy the thesis of the theorem. Moreover, see Figure 2.2, the approximated solution $x^*(s)$ lies within the existence domain of solution obtained above. Finally, remember that $x^*(s)$ is a solution that is beyond the scope of the Newton-Kantorovich theorem.

n	$x_n(s)$	$\|[\mathcal{F}(x_n)](s)\|$
0	s	$7.8009\ldots \times 10^{-2}$
1	$(0.09229591617218\ldots)\,e^{-s} + s$	$1.5047\ldots \times 10^{-3}$
2	$(0.09414952869992\ldots)\,e^{-s} + s$	$6.6518\ldots \times 10^{-7}$
3	$(0.09415034880968\ldots)\,e^{-s} + s$	$1.3042\ldots \times 10^{-13}$
4	$(0.09415034880984\ldots)\,e^{-s} + s$	

Table 2.4: Approximated solution $x^*(s)$ of (2.20) and $\{\|[\mathcal{F}(x_n)](s)\|\}$

n	$\|x^*(s) - x_n(s)\|$	$|t^* - t_n|$
0	$9.4150\ldots \times 10^{-2}$	$1.2380\ldots \times 10^{-1}$
1	$1.8544\ldots \times 10^{-3}$	$1.8787\ldots \times 10^{-2}$
2	$8.2010\ldots \times 10^{-7}$	$6.3325\ldots \times 10^{-4}$
3	$1.6079\ldots \times 10^{-13}$	$7.7713\ldots \times 10^{-7}$

Table 2.5: Absolute errors and error bounds

n	d_n	$t^* - t_n$	e_n
0	$1.2380\ldots \times 10^{-1}$	$1.2380\ldots \times 10^{-1}$	$1.2380\ldots \times 10^{-1}$
1	$1.8787\ldots \times 10^{-2}$	$1.8787\ldots \times 10^{-2}$	$1.9368\ldots \times 10^{-2}$
2	$6.7298\ldots \times 10^{-4}$	$6.3325\ldots \times 10^{-4}$	$3.6365\ldots \times 10^{-4}$
3	$7.1505\ldots \times 10^{-7}$	$7.7713\ldots \times 10^{-7}$	$8.7762\ldots \times 10^{-7}$

Table 2.6: A priori error bounds (by Theorem 2.18)

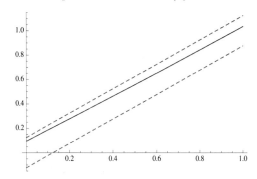

Figure 2.2: Graph (the solid line) of the approximated solution $x^*(s)$ of (2.20)

Example 2.21. We now consider a nonlinear integral equation of Hammerstein-type defined
in (1.36)-(1.39) with $a = 0$, $b = 1$, $u(s) = \frac{2}{3}$, $\mathcal{K}(s,t) = e^{-(s+t)}$ and $\mathcal{H}(x(t)) = \frac{1}{5}x(t)^{7/3} + \frac{1}{4}x(t)^3$;
i.e.:

$$x(s) = \frac{2}{3} + \int_0^1 e^{-(s+t)}\left(\frac{1}{5}x(t)^{7/3} + \frac{1}{4}x(t)^3\right)dt, \quad s \in [0,1]. \tag{2.21}$$

Solving equation (2.21) is equivalent to solving $\mathcal{F}(x) = 0$, where $\mathcal{F} : \mathcal{C}([0,1]) \longrightarrow \mathcal{C}([0,1])$,
is such that

$$[\mathcal{F}(x)](s) = x(s) - \frac{2}{3} - \int_0^1 e^{-(s+t)}\left(\frac{1}{5}x(t)^{7/3} + \frac{1}{4}x(t)^3\right)dt.$$

Moreover,

$$[\mathcal{F}'(x)y](s) = y(s) - \int_0^1 e^{-(s+t)}\left(\frac{7}{15}x(t)^{4/3} + \frac{3}{4}x(t)^2\right)y(t)\,dt,$$

$$[\mathcal{F}''(x)(yz)](s) = \int_0^1 e^{-(s+t)}\left(\frac{7}{45}x(t)^{1/3} + \frac{3}{2}x(t)\right)z(t)y(t)\,dt$$

and

$$\|\mathcal{F}''(x) - \mathcal{F}''(y)\| \leq \frac{e-1}{90\,e}\left(56\|x - y\|^{1/3} + 135\|x - y\|\right) = \omega(\|x - y\|).$$

Observe that the solution $x^*(s)$ of equation (2.21) must satisfy

$$\|x^*(s)\| - \frac{2}{3} - \frac{e-1}{e}\|\left(\frac{1}{5}\|x^*(s)\|^{7/3} + \frac{1}{4}\|x^*(s)\|^3\right) \leq 0,$$

which is true provided that $\|x^*(s)\| \leq \rho_1 = 0.8505\ldots$ or $\|x^*(s)\| \geq \rho_2 = 1.4514\ldots$, where ρ_1
and ρ_2 are the two real positive roots of the scalar equation deduced from the last expression
and given by $\chi(t) = 4 + 6t\,\frac{e-1}{20\,e}\left(4t^{7/3} + 5t^3\right) + t - \frac{2}{3} = 0$.

According to the Newton-Kantorovich theorem, we can approximate, by Newton's method, solutions $x^*(s)$ such that $\|x^*(s)\| \in [0, \rho_1]$, since we can consider the domain

$$\Omega = \{x(s) \in \mathcal{C}([0,1]) : \|x(s)\| < \rho, \, s \in [0,1]\},$$

with $\rho \in (\rho_1, \rho_2)$, as domain for the operator F, where $\|\mathcal{F}''(x)\|$ is bounded, and choose a starting point $x_0(s) \in \Omega$ such that condition (A4) is satisfied. However, the Newton-Kantorovich theorem cannot guarantee the convergence of Newton's method to solutions $x^{**}(s)$ such that $\|x^{**}(s)\| \geq \rho_2$, because we cannot choose a domain where $x^{**}(s)$ lies, since if the domain is chosen at random, it could not contain $x^{**}(s)$ or cut it, and in such case $\|\mathcal{F}''(x)\|$ is not bounded. We see in this example that both situations are covered by Theorem 2.16 and in the first one, where the Newton-Kantorovich theorem also covers it, we improve the domains of existence and uniqueness of solution and the error bounds that are obtained from the Newton-Kantorovich theorem.

We begin with the case $\|x^*(s)\| \in [0, \rho_1]$ and $\Omega = \{x(s) \in \mathcal{C}([0,1]) : \|x(s)\| < \rho, \, s \in [0,1]\}$, with $\rho \in (\rho_1, \rho_2)$. For example, we choose $\rho = \frac{5}{4}$ and $x_0(s) = \frac{2}{3}$ as starting point. So, $\beta = 1.5180\ldots$, $\eta = 0.1455\ldots$, $\delta = 0.9757\ldots$, $M = 1.6089\ldots$ and $M\beta\eta = 0.3555\ldots \leq \frac{1}{2}$. Therefore, from the Newton-Kantorovich theorem, the semilocal convergence of Newton's method to a solution $x^*(s)$ in $\Omega = \{x(s) \in \mathcal{C}([0,1]) : \|x(s)\| < \frac{5}{4}, \, s \in [0,1]\}$ is guaranteed. As Kantorovich's polynomial is $p(s) = (0.8044\ldots)s^2 - (0.6587\ldots)s + (0.0959\ldots)$, the real positive roots are $s^* = 0.1893\ldots$ and $s^{**} = 0.6294\ldots$ and $B(x_0, s^*) \subseteq \Omega$, we obtain that the domains of existence and uniqueness of solutions are respectively

$$\{\nu \in \Omega : \|\nu(s) - x_0(s)\| \leq 0.1893\ldots\} \quad \text{and} \quad \{\nu \in \Omega : \|\nu(s) - x_0(s)\| < 0.6294\ldots\}.$$

On the other hand, we choose $w_0(t) = w(t) = \frac{e-1}{90\,e}\left(56\sqrt[3]{t} + 135t\right)$ for Theorem 2.16 and then

$$\varphi(t) = (0.0959\ldots) - (0.6587\ldots)t + (0.4878\ldots)t^2 + (0.1264\ldots)t^{7/3} + (0.1580\ldots)t^3,$$

with $t_0 = 0$, has two positive real zeros are $t^* = 0.1718\ldots$ and $t^{**} = 0.7356\ldots$, so that the domains of existence and uniqueness of solutions are respectively

$$\{\nu \in \mathcal{C}([0,1]) : \|\nu(s) - x_0(s)\| \leq 0.1718\ldots\} \text{ and } \{\nu \in \mathcal{C}([0,1]) : \|\nu(s) - x_0(s)\| < 0.7356\ldots\}.$$

Therefore, the domains obtained from the Newton-Kantorovich theorem are improved, since the domain of existence of solution is smaller and the domain of uniqueness of solution is bigger.

Once the convergence of Newton's method is guaranteed, we approximate a solution $x^*(s)$ by the method. For this, we proceed as in Section 1.2 to construct the iterations of Newton's method. Then, after four iterations of Newton's method, we obtain the approximated solution $x^*(s) = (0.1362\ldots)e^{-s} + 2/3$ given in Table 2.7. From the sequence $\{\|[\mathcal{F}(x_n)](s)\|\}$ given in Table 2.7, we observe that $x^*(s)$ is a good approximation of a solution of (2.21). In Table 2.8, we show the errors $\|x^*(s) - x_n(s)\|$ and the error bounds. Moreover, see Figure 2.3, the approximated solution $x^*(s)$ lies within the existence domain of solution obtained above and observe that $\|x^*(s)\| = 0.8029\ldots \leq \rho_1 = 0.8505\ldots$

In Table 2.8, we can see that the error bounds $\{|t^* - t_n|\}$ obtained from Theorem 2.16 improves those, $\{|s^* - s_n|\}$, given by the majorizing real sequence obtained from the previous Kantorovich polynomial $p(s)$.

n	$x_n(s)$	$\|[\mathcal{F}(x_n)](s)\|$
0	$2/3$	$9.5908\ldots \times 10^{-2}$
1	$(0.12988929787142\ldots)\,\mathrm{e}^{-s}+2/3$	$4.2824\ldots \times 10^{-3}$
2	$(0.13626918161654\ldots)\,\mathrm{e}^{-s}+2/3$	$1.1109\ldots \times 10^{-5}$
3	$(0.13628581864697\ldots)\,\mathrm{e}^{-s}+2/3$	$7.5799\ldots \times 10^{-11}$
4	$(0.13628581876049\ldots)\,\mathrm{e}^{-s}+2/3$	

Table 2.7: Approximated solution $x^*(s)$ of (2.21) and $\{\|[\mathcal{F}(x_n)](s)\|\}$

n	$\|x^*(s)-x_n(s)\|$	$|t^*-t_n|$	$|s^*-s_n|$
0	$1.3628\ldots \times 10^{-1}$	$1.7182\ldots \times 10^{-1}$	$1.8939\ldots \times 10^{-1}$
1	$6.3965\ldots \times 10^{-3}$	$2.6231\ldots \times 10^{-2}$	$4.3806\ldots \times 10^{-2}$
2	$1.6637\ldots \times 10^{-5}$	$9.4743\ldots \times 10^{-4}$	$3.6365\ldots \times 10^{-3}$
3	$1.1351\ldots \times 10^{-10}$	$1.3522\ldots \times 10^{-6}$	$2.9562\ldots \times 10^{-5}$

Table 2.8: Absolute errors and error bounds

After that, we remember that we have seen previously that equation (2.21) may have a solution $x^{**}(s)$ such that $\|x^{**}(s)\| \geq \rho_2 = 1.4514\ldots$ In this case, we guarantee the semilocal convergence of Newton's method from Theorem 2.16. In this case, $\Omega = \mathcal{C}([0,1])$. In view of how it is the solution $x^*(s)$ of equation (2.21) and taking into account that $\|x^{**}(s)\| \geq \rho_2 = 1.4514\ldots$, a reasonable choice of initial approximation seems to be $x_0(s) = 2\,\mathrm{e}^{-s}+\frac{2}{3}$, that satisfies $\|x_0(s)\| = \frac{8}{3} > \rho_2 = 1.4514\ldots$ At first, we cannot apply Theorem 2.16, since the corresponding majorant function

$$\varphi(t) = (0.1556\ldots) - (0.2772\ldots)t + (1.5369\ldots)t^2 + (0.1264\ldots)t^{7/3} + (0.1580\ldots)t^3,$$

with $t_0 = 0$, does not have any real positive root. However, it seems clear that improving the initial approximation, the conditions of Theorem 2.16 hold. Indeed, applying Newton's method from $x_0(s) = 2\,\mathrm{e}^{-s}+\frac{2}{3}$, after two iterations, we obtain the approximation $x_2(s) = (1.8392\ldots)\,\mathrm{e}^{-s}+\frac{2}{3}$, that satisfies, as new starting point, the conditions of Theorem 2.16, since

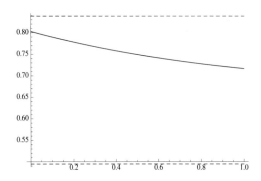

Figure 2.3: Graph (the solid line) of the approximated solution $x^*(s)$ of (2.21)

if we take $\tilde{x}_0(s) = x_2(s)$, we obtain that the new majorant function for Theorem 2.16 is

$$\varphi(t) = (0.0001\ldots) - (0.2556\ldots)t + (1.4551\ldots)t^2 + (0.1264\ldots)t^{7/3} + (0.1580\ldots)t^3,$$

with $t_0 = 0$, and has two positive zeros $t^* = 0.0005\ldots$ and $t^{**} = 0.1642\ldots$, so that the domains of existence and uniqueness of solutions are respectively

$$\{\nu \in C([0,1]) : \|\nu(s) - \tilde{x}_0(s)\| \leq 0.0005\ldots\} \text{ and } \{\nu \in C([0,1]) : \|\nu(s) - \tilde{x}_0(s)\| < 0.1642\ldots\}.$$

Observe that the new starting point $\tilde{x}_0(s)$ is such that $\|\tilde{x}_0(s)\| = 2.5059\ldots > \rho_2 = 1.4514\ldots$ After three more iterations of Newton's method, we obtain the approximated solution $x^{**}(s) = (1.8390\ldots) e^{-s} + 2/3$ given in Table 2.9. Observe, from the sequence $\{\|[\mathcal{F}(\tilde{x}_n)](s)\|\}$ given in Table 2.9, that $x^{**}(s)$ is a good approximation of a solution of (2.21). In Table 2.10, we show the errors $\|x^{**}(s) - \tilde{x}_n(s)\|$ and the error bounds. Moreover, observe that $\|x^{**}(s)\| = 2.5057\ldots > \rho_2 = 1.4514\ldots$, so that $x^{**}(s)$ is a solution that is beyond the scope of the Newton-Kantorovich theorem.

Finally, analogous conclusions to the previous example about the a priori error bounds obtained from Theorem 2.18 can be drawn.

n	$\tilde{x}_n(s)$	$\|[\mathcal{F}(\tilde{x}_n)](s)\|$
0	$(1.839266741339092\ldots) e^{-s} + 2/3$	$1.4811\ldots \times 10^{-4}$
1	$(1.839095613517311\ldots) e^{-s} + 2/3$	$1.8233\ldots \times 10^{-8}$
2	$(1.839095592446375\ldots) e^{-s} + 2/3$	$2.7642\ldots \times 10^{-16}$
3	$(1.839095592446374\ldots) e^{-s} + 2/3$	

Table 2.9: Approximated solution $x^{**}(s)$ of (2.21) and $\{\|[\mathcal{F}(\tilde{x}_n)](s)\|\}$

n	$\|x^{**}(s) - \tilde{x}_n(s)\|$	$|t^* - t_n|$
0	$1.7114\ldots \times 10^{-4}$	$5.8138\ldots \times 10^{-4}$
1	$2.1070\ldots \times 10^{-8}$	$1.9382\ldots \times 10^{-6}$
2	$3.1944\ldots \times 10^{-16}$	$2.1776\ldots \times 10^{-11}$

Table 2.10: Absolute errors and error bounds

2.1.3.6 A little more: three particular cases

Case 1. Taking into account Remark 2.4, where we note that it is enough to satisfy condition (2.9),

$$\|F''(x) - F''(x_0)\| \leq K_0 \|x - x_0\|, \quad \text{for} \quad x \in \Omega,$$

instead of condition (B2) to prove the Huang theorem, and following an analogous procedure to that given in Remark 2.5, we obtain, by solving the problem of interpolation fitting given by conditions (B1) and (2.9), the function

$$\phi_0(t) = \frac{K_0}{6} t^3 + \frac{\delta}{2} t^2 - \frac{t}{\beta} + \frac{\eta}{\beta}.$$

Observe also that we can obtain, similarly to what has been done above for function (2.18), function ϕ_0 from the resolution of an initial value problem. Indeed, if we try to construct a majorant sequence from a real function f, we obtain from (D2) that

$$\|F''(x) - F''(x_0)\| \le K_0\|x - x_0\| \le K_0(t - t_0)$$

if $\|x - x_0\| \le t - t_0$ and look for a real function f such that $K_0(t - t_0) = f''(t) - f''(t_0)$. So,

$$\int_{t_0}^{t} f'''(\xi)\, d\xi = K_0 \int_{t_0}^{t} d\xi$$

and, taking into account conditions (B1)-(2.9), we can then obtain the function

$$\tilde{\phi}_0(t) = \frac{K_0}{6}(t - t_0)^3 + \frac{\delta}{2}(t - t_0)^2 - \frac{t - t_0}{\beta} + \frac{\eta}{\beta},$$

for a fixed $t_0 \ge 0$, as the unique solution of the initial value problem

$$\begin{cases} y'''(t) - K_0 = 0, \\ y(t_0) = \dfrac{\eta}{\beta}, \quad y'(t_0) = -\dfrac{1}{\beta}, \quad y''(t_0) = \delta, \end{cases}$$

since $\tilde{\phi}_0'''(t) = K_0$, $\tilde{\phi}_0''(t_0) = \delta$, $-\frac{1}{\tilde{\phi}_0'(t_0)} = \beta$ and $-\frac{\tilde{\phi}_0(t_0)}{\tilde{\phi}_0'(t_0)} = \eta$. Observe that the function $\tilde{\phi}_0$ also has the property of the translation of Kantorovich's polynomial (1.23) given in Remark 1.18, so that it majorizes, independent of the value of t_0, the operator F in the sense of Definition 1.26. So, if $t_0 = 0$, then the function $\tilde{\phi}_0(t)$ is reduced to

$$\phi_0(t) = \frac{K_0}{6}t^3 + \frac{\delta}{2}t^2 - \frac{t}{\beta} + \frac{\eta}{\beta}.$$

In addition, we can also see that Huang's polynomial $\phi(t)$ given in (2.2) is obtained from an initial value problem. In particular, we obtain

$$\tilde{\phi}(t) = \frac{K}{6}(t - t_0)^3 + \frac{\delta}{2}(t - t_0)^2 - \frac{t - t_0}{\beta} + \frac{\eta}{\beta}$$

as the unique solution of the initial value problem

$$\begin{cases} y'''(t) - K = 0, \\ y(t_0) = \dfrac{\eta}{\beta}, \quad y'(t_0) = -\dfrac{1}{\beta}, \quad y''(t_0) = \delta, \end{cases}$$

just keep in mind that

$$\|F''(x) - F''(y)\| \le K\|x - y\| \le K(u - v) = f'''(u) - f'''(v)$$

if $\|x - y\| \le u - v$, and

$$\int_{u}^{v} K\, d\tau = \int_{u}^{v} f'''(\tau)\, d\tau.$$

Moreover, the function $\tilde{\phi}(t)$ is reduced to $\phi(t)$ if $t_0 = 0$. Later, we see the relationship between the two situations deduced from considering ϕ and ϕ_0.

If we suppose that the operator F'' is Lipschitz continuous in Ω, then F'' satisfies conditions (B2) and (2.9), so that $K_0 \leq K$ and $\phi_0(t) \leq \phi(t)$, for $t \geq 0$, and Theorem 2.16 has an additional interest, since we can obtain better domains of existence and uniqueness of solution and tighter error bounds, as we can see in the next example. Highlight that these improvements rely on the idea, which is observed both in Kantorovich's polynomial, (1.23), and Huang's polynomial, (2.2), that majorant function depends on the starting point. That is why, in our subsequent analyses, we consider the general condition required for the second derivative of the operator involved in the starting point instead of the all domain Ω.

Example 2.22. We consider a nonlinear integral equation of Hammerstein-type defined in (1.36)-(1.39) with $a = -\frac{1}{2}$, $b = \frac{1}{2}$, $u(s) = s$, $\mathcal{K}(s,t) = e^{st}$ and $\mathcal{H}(x(t)) = \varepsilon x(t)^4$; i.e.:

$$x(s) = s + \varepsilon \int_{-\frac{1}{2}}^{\frac{1}{2}} e^{st} x(t)^4 \, dt, \quad \varepsilon = \frac{3}{4}, \quad s \in [-\tfrac{1}{2}, \tfrac{1}{2}]. \tag{2.22}$$

Solving equation (2.22) is equivalent to solving $\mathcal{F}(x) = 0$, where $\mathcal{F} : \mathcal{C}([-\tfrac{1}{2}, \tfrac{1}{2}]) \longrightarrow \mathcal{C}([-\tfrac{1}{2}, \tfrac{1}{2}])$ and

$$[\mathcal{F}(x)](s) = x(s) - s - \varepsilon \int_{-\frac{1}{2}}^{\frac{1}{2}} e^{st} x(t)^4 \, dt.$$

Moreover,

$$[\mathcal{F}'(x)y](s) = y(s) - 4\varepsilon \int_{-\frac{1}{2}}^{\frac{1}{2}} e^{st} x(t)^3 y(t) \, dt, \tag{2.23}$$

$$[\mathcal{F}''(x)(yz)](s) = -12\varepsilon \int_{-\frac{1}{2}}^{\frac{1}{2}} e^{st} x(t)^2 z(t) y(t) \, dt$$

and, since $\varepsilon = \frac{3}{4}$,

$$\|\mathcal{F}''(x) - \mathcal{F}''(y)\| \leq 9(1.0104\ldots)\,(\|x\| + \|y\|)\,\|x - y\|.$$

Note again that condition (A2) of the Newton-Kantorovich theorem is not satisfied in general because $\|\mathcal{F}''(x)\|$ is not bounded in $\mathcal{C}([-\tfrac{1}{2}, \tfrac{1}{2}])$.

We now observe that the solution $x^*(s)$ must satisfy that $\|x^*(s)\| \leq \rho_1 = 0.5949\ldots$ or $\|x^*(s)\| \geq \rho_2 = 0.7788\ldots$, where ρ_1 and ρ_2 are the two real positive roots of the scalar equation $\chi(t) = 3 \sinh\left(\frac{1}{4}\right) t^4 - t + \frac{1}{2} = 0$, which is deduced from (1.38). We then consider $\Omega = \{x(s) \in \mathcal{C}([-\tfrac{1}{2}, \tfrac{1}{2}]) : \|x(s)\| < 2, \, s \in [0, 1]\}$ and choose $x_0(s) = s \in \Omega$ as the starting point for Newton's method and Theorem 2.16. As a consequence, $\beta = 1.6100\ldots$, $\eta = 0.0762\ldots$, $\delta = 2.2735\ldots$, $K = 36.3762\ldots$ and $K_0 = 22.7351\ldots$, so that

$$\phi(t) = (6.0627\ldots)t^3 + (1.1367\ldots)t^2 - (0.6210\ldots)t + (0.0473\ldots),$$

whose two positive real roots are $r^* = 0.1167\ldots$ and $r^{**} = 0.1480\ldots$, and

$$\phi_0(t) = (3.7891\ldots)t^3 + (1.1367\ldots)t^2 - (0.6210\ldots)t + (0.0473\ldots),$$

whose two positive real roots are $t^* = 0.1014\ldots$ and $t^{**} = 0.2035\ldots$ Observe now that $K_0 \leq K$ and $\phi_0(t) \leq \phi(t)$, see Figure 2.4. Therefore, we obtain the domains of existence and uniqueness of solutions of equation (2.22) are better if $\phi_0(t)$ is considered instead of $\phi(t)$,

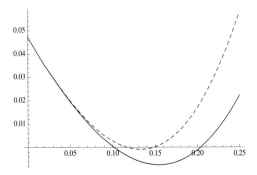

Figure 2.4: Functions $\phi(t)$, the dashed line, and $\phi_0(t)$, the solid line, of Example 2.22

n	$|t^* - t_n|$	$|r^* - r_n|$
0	$1.0149\ldots \times 10^{-1}$	$1.1671\ldots \times 10^{-1}$
1	$2.5232\ldots \times 10^{-2}$	$4.0452\ldots \times 10^{-2}$
2	$3.5026\ldots \times 10^{-3}$	$1.1325\ldots \times 10^{-2}$
3	$9.6061\ldots \times 10^{-5}$	$2.8469\ldots \times 10^{-3}$
4	$7.7206\ldots \times 10^{-8}$	$2.0682\ldots \times 10^{-4}$
5	$4.9968\ldots \times 10^{-14}$	$1.2745\ldots \times 10^{-6}$

Table 2.11: Error bounds from functions $\phi_0(t)$ and $\phi(t)$ of Example 2.22

since $t^* < r^*$ and $t^{**} > r^{**}$. Moreover, we also obtain better error bounds from the majorizing real sequence $\{t_n\}$, where $\{t_n\}$ is obtained when Newton's method is applied to the function $\phi_0(t)$, than from majorizing sequence $\{r_n\}$, where $\{r_n\}$ is obtained when Newton's method is applied to the function $\phi(t)$, as we can see in Table 2.11.

Next, we approximate a solution of the integral equation. For this, we first observe that the usual expansion for $\mathcal{K}(s,t) = e^{st}$ shows that the kernel of integral equation (2.22) is not separable. Then, from (2.23), we define

$$[T'(x)y](s) = y(s) - 4\varepsilon \int_{-\frac{1}{2}}^{\frac{1}{2}} \widetilde{\mathcal{K}}(s,t)x(t)^3 y(t)\, dt,$$

where $\widetilde{\mathcal{K}}(s,t)$ is Taylor's series for $\mathcal{K}(s,t) = e^{st}$ when it is developed for the second derivative in $t = 0$, so that

$$\mathcal{K}(s,t) = e^{st} = \widetilde{\mathcal{K}}(s,t) + \mathcal{R}(\theta,s,t); \quad \widetilde{\mathcal{K}}(s,t) = 1 + st + \frac{1}{2}s^2 t^2, \quad \mathcal{R}(\theta,s,t)$$

$$= \frac{e^{s\theta}}{6}s^3 t^3, \ \theta \in (\min\{0,\ t\}, \max\{0,\ t\}).$$

Next, we approximate $[\mathcal{F}'(x)y](s)$ by $[T'(x)y](s)$ and look for $[T'(x)]^{-1}$. For this, if we denote $I_j^x = 4\int_{-\frac{1}{2}}^{\frac{1}{2}} \gamma_j(t)x(t)^3 y(t)\, dt$, we have

$$[T'(x)]^{-1}z(s) = z(s) + \varepsilon \sum_{j=1}^{3} \sigma_j(s)I_j^x, \tag{2.24}$$

where $\gamma_j(t) = t^{j-1}$ and $\sigma_j(t) = \dfrac{t^{j-1}}{(j-1)!}$, for $j = 1, 2, 3$, provided that the integrals I_j^x can be calculated independently of y, what can be done in the following way. Equality (2.24) is first multiplied by $4\gamma_i(s)x(s)^3$ and second integrated in the s variable, so that we obtain the following linear system

$$I_i^x - \varepsilon \sum_{j=1}^{3} a_{ij}(x)I_j^x = b_i(x), \quad i = 1, 2, 3, \tag{2.25}$$

where $a_{ij}(x) = 4\int_{-\frac{1}{2}}^{\frac{1}{2}} \gamma_i(s)x(s)^3\sigma_j(s)\, ds$ and $b_i(x) = 4\int_{-\frac{1}{2}}^{\frac{1}{2}} \gamma_i(s)x(s)^3 z(s)\, ds$. System (2.25) has a unique solution if

$$(-\varepsilon)^3 \begin{vmatrix} a_{11}(x) - \frac{1}{\varepsilon} & a_{12}(x) & a_{13}(x) \\ a_{21}(x) & a_{22}(x) - \frac{1}{\varepsilon} & a_{23}(x) \\ a_{31}(x) & a_{32}(x) & a_{33}(x) - \frac{1}{\varepsilon} \end{vmatrix} \neq 0.$$

Now, if we assume that $\frac{1}{\varepsilon}$ is not an eigenvalue of the matrix $(a_{ij}(x))_{i,j=1,2,3}$, then I_j^x, for $j = 1, 2, 3$, is the solution of system (2.25) and can be then define

$$[\mathcal{F}'(x)]^{-1}z(s) = z(s) + \varepsilon \sum_{j=1}^{3} \sigma_j(s)I_j^x,$$

In addition, we can then apply Newton's method.

Note that the last condition required for ε has to be satisfied in each iteration of Newton's method.

To calculate the iterations $x_n(s)$ of Newton's method from the starting point $x_0(s) = s$ and $\varepsilon = \frac{3}{4}$, we do the following. First, we solve system (2.25) when $x(s)$ is $x_0(s) = s$. Second, we calculate

$$x_1(s) = x_0(s) - F(x_0)(s) - \varepsilon \sum_{j=1}^{3} \sigma_j(s)I_j^x$$
$$= (0.009388\ldots) + (1.000357\ldots)s + (0.000838\ldots)s^2.$$

Third, we repeat the last and, after three iterations of Newton's method, we obtain the approximated solution $x^*(s) = (0.009422\ldots) + (1.000359\ldots)s + (0.000840\ldots)s^2$ given in Table 2.12. From the sequence $\{\|[\mathcal{F}(x_n)](s)\|\}$ given in Table 2.12, we observe that $x^*(s)$ is a good approximation of a solution of (2.22). In Table 2.13, we show the errors $\|x^*(s) - x_n(s)\|$ and the error bounds. Moreover, see Figure 2.5, the approximated solution $x^*(s)$ lies within the existence domain of solution obtained above.

Case 2. Two particular cases of condition (U2) are (2.9) and the following one, that is presented in [37], where Gutiérrez assumes that F'' satisfies in Ω the following Hölder-type condition:

$$\|F''(x) - F''(x_0)\| \le K_0\|x - x_0\|^p, \quad p \ge 0, \quad \text{for} \quad x \in \Omega. \tag{2.26}$$

Note that condition (2.26) is also reduced to condition (2.9) if $p = 1$. We then say that F'' satisfies a center Hölder condition or a center Lipschitz condition, respectively. Both cases are particular cases of condition (U2) if $\omega_0(z) = K_0z^p$ or $\omega_0(z) = K_0z$, respectively.

In particular, Gutiérrez gives a new semilocal convergence result (Theorem 2.23, see below) for Newton's method under the following conditions:

n	$x_n(s)$	$\|[\mathcal{F}(x_n)](s)\|$
0	s	$1.0212\ldots \times 10^{-2}$
1	$(0.009388412954\ldots) + (1.000357678774\ldots)s + (0.000838251156\ldots)s^2$	$3.7144\ldots \times 10^{-5}$
2	$(0.009422715820\ldots) + (1.000359587599\ldots)s + (0.000840837701\ldots)s^2$	$5.0347\ldots \times 10^{-10}$
3	$(0.009422716278\ldots) + (1.0003595876332\ldots)s + (0.000840837735\ldots)s^2$	

Table 2.12: Approximated solution $x^*(s)$ of (2.22) and $\{\|[\mathcal{F}(x_n)](s)\|\}$

n	$\|x^*(s) - x_n(s)\|$	$|t^* - t_n|$
0	$1.0623\ldots \times 10^{-2}$	$1.0149\ldots \times 10^{-1}$
1	$3.8798\ldots \times 10^{-5}$	$2.5232\ldots \times 10^{-2}$
2	$5.2590\ldots \times 10^{-10}$	$3.5026\ldots \times 10^{-3}$

Table 2.13: Absolute errors and error bounds

(C1) There exists $\Gamma_0 = [F'(x_0)]^{-1} \in \mathcal{L}(Y, X)$, for some $x_0 \in \Omega$, with $\|\Gamma_0\| \leq \beta$ and $\|\Gamma_0 F(x_0)\| \leq \eta$; moreover, $\|F''(x_0)\| \leq \delta$.

(C2) There exist two constants $K \geq 0$ and $p \geq 0$ such that $\|F''(x) - F''(x_0)\| \leq K_0\|x - x_0\|^p$, for $x \in \Omega$.

(C3) $\psi(\alpha) \leq 0$, where ψ is the function

$$\psi(t) = \frac{K_0}{(p+1)(p+2)}t^{2+p} + \frac{\delta}{2}t^2 - \frac{t}{\beta} + \frac{\eta}{\beta} \qquad (2.27)$$

and α is the unique positive solution of $\psi'(t) = 0$, and $B(x_0, t^*) \subset \Omega$, where t^* is the smallest positive solution of $\psi(t) = 0$.

Observe that function (2.18) is reduced to function (2.27) if condition (U2) is reduced to condition (C2) and choose $t_0 = 0$.

Notice that Remark 2.2 is also satisfied when the function ϕ is substituted by the function ψ. In this case, as a particular case of Theorem 2.16, we obtain the following result given by Gutiérrez in [37].

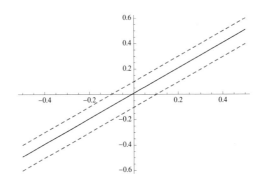

Figure 2.5: Graph (the solid line) of the approximated solution $x^*(s)$ of (2.22)

Theorem 2.23. *Let $F : \Omega \subseteq X \longrightarrow Y$ be a twice continuously Fréchet differentiable operator defined on a nonempty open convex domain Ω of a Banach space X with values in a Banach space Y. Suppose that conditions (C1)-(C2)-(C3) are satisfied. Then, Newton's sequence, given by (1.1) and starting at x_0, converges to a solution x^* of $F(x) = 0$. Moreover, $x_n, x^* \in B(x_0, t^*)$, for all $n \in \mathbb{N}$, and x^* is unique in $B(x_0, t^{**}) \cap \Omega$ if $t^* < t^{**}$ or in $\overline{B(x_0, t^*)}$ if $t^* = t^{**}$. Furthermore,*

$$\|x^* - x_n\| \leq t^* - t_n, \quad n = 0, 1, 2, \ldots,$$

where $t_n = t_{n-1} - \frac{\psi(t_{n-1})}{\psi'(t_{n-1})}$, with $n \in \mathbb{N}$, $t_0 = 0$ and $\psi(t)$ is defined in (2.27).

Moreover, the error estimates given in Theorem 2.18 are reduced to those given in [37] for Newton's method when Newton's real sequence is defined from functions (2.2) and (2.27), which are particular cases of function (2.18).

Case 3. We have just seen in Example 2.22 that $\mathcal{F}''$ is Lipschitz continuous in a domain and, therefore, in particular, $\mathcal{F}''$ is Lipschitz continuous in one point of the domain. Now, we see that there are situations in which $\mathcal{F}''$ is only Lipschitz continuous in some points of a domain, but not in all the domain, what justifies the study carried out previously for center conditions. For example, the following nonlinear integral equation of the Hammerstein-type and form (1.36)-(1.39):

$$x(s) = u(s) + \lambda \int_a^b \mathcal{K}(s,t) \, x(t)^{2+\frac{1}{n}} \, dt, \quad s \in [a,b], \quad \lambda \in \mathbb{R}, \quad n \text{ is odd}, \tag{2.28}$$

where u is such that $u(s) > 0$, the kernel $\mathcal{K}$ is continuous and positive in $[a,b] \times [a,b]$. Integral equations of this kind are treated in [13, 75].

Solving equation (2.28) is equivalent to solving the equation $\mathcal{F}(x) = 0$, where $\mathcal{F} : \Omega \subseteq \mathcal{C}([a,b]) \longrightarrow \mathcal{C}([a,b])$ and

$$[\mathcal{F}(x)](s) = x(s) - u(s) - \lambda \int_a^b \mathcal{K}(s,t) \, x(t)^{2+1/n} \, dt, \quad s \in [a,b], \quad \lambda \in \mathbb{R}, \quad n \text{ is odd}.$$

In this case, we consider $\Omega = \mathcal{C}([a,b])$ and have

$$[\mathcal{F}'(x)y](s) = y(s) - \lambda \left(2 + \frac{1}{n}\right) \int_a^b \mathcal{K}(s,t) \, x(t)^{1+1/n} \, y(t) \, dt,$$

$$[\mathcal{F}''(x)(yz)](s) = -\lambda \left(2 + \frac{1}{n}\right)\left(1 + \frac{1}{n}\right) \int_a^b \mathcal{K}(s,t) \, x(t)^{1/n} \, z(t)y(t) \, dt.$$

We then note that $\mathcal{F}''$ is not Lipschitz continuous in all domain Ω. For this, we consider $[a,b] = [0,1]$, $\mathcal{K}(s,t) = 1$ and $\ell(t) = 0$. Then, $[\mathcal{F}''(\ell)(yz)](s) = 0$ and

$$\|\mathcal{F}''(x) - \mathcal{F}''(\ell)\| = |\lambda| \left(2 + \frac{1}{n}\right)\left(1 + \frac{1}{n}\right) \int_0^1 x(t)^{\frac{1}{n}} \, dt.$$

If $\mathcal{F}''$ were Lipschitz continuous in Ω, then

$$\|\mathcal{F}''(x) - \mathcal{F}''(\ell)\| \leq K_1 \|x - \ell\|,$$

or equivalently, the inequality

$$\int_0^1 x(t)^{\frac{1}{n}} \, dt \leq K_2 \max_{s \in [0,1]} |x(s)|, \tag{2.29}$$

were satisfied for all $x \in \Omega$ and a constant K_2. But this is not true, as we can see in the following. Indeed, if we consider the functions

$$x_i(t) = \frac{t}{i}, \quad i \geq 1, \quad t \in [0, 1],$$

and they are substituted into (2.29),

$$\frac{1}{i^{\frac{1}{n}}\left(1 + \frac{1}{n}\right)} \leq \frac{K_2}{i} \quad \Longleftrightarrow \quad i^{1-\frac{1}{n}} \leq K_2\left(1 + \frac{1}{n}\right), \quad \text{for all } i \geq 1,$$

then inequality (2.29) is not true when $i \to +\infty$.

However, $\mathcal{F}''$ is Lipschitz continuous in a point of Ω. For this, we consider $x_0(t) = u(t)$ and $d = \min_{s \in [a,b]} u(s)$, $d > 0$. Then, for $y, z \in \Omega$, we have

$$\|[\mathcal{F}''(x) - \mathcal{F}''(x_0)](yz)\| = |\lambda|\left(2 + \frac{1}{n}\right)\left(1 + \frac{1}{n}\right) \max_{s \in [a,b]} \left| \int_a^b \mathcal{K}(s,t)\left(x(t)^{\frac{1}{n}} - u(t)^{\frac{1}{n}}\right) z(t)y(t)\, dt \right|$$

$$\leq |\lambda|\left(2 + \frac{1}{n}\right)\left(1 + \frac{1}{n}\right)$$

$$\times \max_{s \in [a,b]} \int_a^b \frac{\mathcal{K}(s,t)\|x(t) - u(t)\|}{x(t)^{\frac{n-1}{n}} + x(t)^{\frac{n-2}{n}}u(t)^{\frac{1}{n}} + \cdots + u(t)^{\frac{n-1}{n}}}\, dt \, \|z\|\|y\|,$$

so that

$$\|\mathcal{F}''(x) - \mathcal{F}''(x_0)\| \leq \frac{|\lambda|\left(2 + \frac{1}{n}\right)\left(1 + \frac{1}{n}\right)}{d^{\frac{n-1}{n}}}\left(\max_{s \in [a,b]} \int_a^b \mathcal{K}(s,t)\, dt\right)\|x - x_0\| \leq K\|x - x_0\|,$$

where $K = \dfrac{|\lambda|\left(2 + \frac{1}{n}\right)\left(1 + \frac{1}{n}\right)}{d^{\frac{n-1}{n}}}\left(\max\limits_{s \in [a,b]} \int_a^b \mathcal{K}(s,t)\, dt\right)$. As a consequence, $\mathcal{F}''$ is Lipschitz continuous in $x_0 \in \Omega$.

Finally, we note that if n is even, then we choose $\Omega = \{x(s) \in \mathcal{C}([a,b]) : x(s) \geq 0, s \in [a,b]\}$.

An important consequence of the above-mentioned is that the semilocal convergence results for Newton's method under center conditions for F'' are interesting regardless of the general situation is not centered.

2.2 Operators with ω-bounded second-derivative

Remember that condition (A2) of Kantorovich requires that the second derivative of the operator involved, F'', is bounded in a domain Ω, where a solution x^* must exist. According to this and Remark 1.32, the number of equations that can be solved by Newton's method is limited, since it is not easy to see that F'' is bounded in a general domain Ω. It is not easy either to locate a domain where F'' is bounded and the solution x^* is contained.

In Section 1.2, we have seen that integral equation (1.40) can have a solution $x^{**}(s)$ such that $\|x^{**}(s)\| \geq \rho_2 = 4.1853\ldots$, but the convergence of Newton's method cannot be guaranteed from the Newton-Kantorovich theorem, since a domain where $x^{**}(s)$ lies and $\|F''(x)\|$ is bounded cannot be fixed. However, the convergence of Newton's method can be guaranteed from Theorem 1.27 if we can determine a suitable majorant function f. So, our immediate aim will be sought for scalar majorant functions of the operator involved that permit to guarantee the convergence of Newton's method in situations in which the Newton-Kantorovich theorem cannot, as for example the situation we have just indicated above.

From now on, our interest is focused on generalize the semilocal convergence conditions given by Kantorovich for Newton's method in the Newton-Kantorovich theorem, so that condition (A2) is relaxed in order to Newton's method can be applied to solve more equations. In Section 1.2, we have seen that condition (A2) limits the application of the Newton-Kantorovich theorem. To improve what Kantorovich does, we replace condition (A2) with the following milder condition

$$\|F''(x)\| \leq \omega(\|x\|), \quad x \in \Omega, \tag{2.30}$$

where $\omega : [0, +\infty) \longrightarrow \mathbb{R}$ is a continuous monotonous (nondecreasing or nonincreasing) function such that $\omega(0) \geq 0$. Obviously, condition (2.30) generalizes condition (A2).

To do the last generalization, we follow the modification of the majorant principle of Kantorovich introduced in Section 1.1.3. Remember that if f is a majorant function of the operator F, it is sufficient to guarantee the convergence of the scalar sequence $\{t_{n+1} = N_f(t_n)\}$, with t_0 given, to have guaranteed the convergence of Newton's method, $\{x_{n+1} = N_F(x_n)\}$, with x_0 given, in the Banach space X. In particular, we construct majorizing real sequences *ad hoc*, so that these sequences are adapted to particular problems, since the modification of condition (A2) proposed gives more information about the operator F, not just that F'' is bounded, as condition (A2) does.

On the other hand, if we consider (A1)-(2.30), instead of (A1)-(A2), we cannot obtain a real function f by interpolation fitting, as Kantorovich does, since (2.30) does not allow determining the class of functions where (A1) can be applied. To solve this problem, we proceed differently, without interpolation fitting, by solving an initial value problem, as we do at the end of Section 1.1.4 to obtain the majorant function given by Kantorovich's polynomial (1.23).

This way of getting a majorant function has the advantage of being able to be generalized to conditions (A1)-(2.30). To do this, as ω is monotonous, we have

$$\|F''(x)\| \leq \omega(\|x\|) \leq \varpi(t; \|x_0\|, t_0),$$

where

$$\varpi(t; \|x_0\|, t_0) = \begin{cases} \omega(\|x_0\| - t_0 + t) & \text{if } \omega \text{ is nondecreasing, provided that} \\ & \|x\| - \|x_0\| \leq \|x - x_0\| \leq t - t_0, \\ \omega(\|x_0\| + t_0 - t) & \text{if } \omega \text{ is nonincreasing, provided that} \\ & \|x_0\| - \|x\| \leq \|x - x_0\| \leq t - t_0. \end{cases} \tag{2.31}$$

Observe that $\varpi(t; \|x_0\|, t_0)$ is a nondecreasing function.

As a result of the above, instead of (2.30), we consider

$$\|F''(x)\| \leq \varpi(t; \|x_0\|, t_0), \quad \text{for} \quad \|x - x_0\| \leq t - t_0,$$

where $\varpi : [t_0, +\infty) \longrightarrow \mathbb{R}$ is a continuous nondecreasing function such that $\varpi(t_0; \|x_0\|, t_0) \geq 0$.

Now, we obtain the semilocal convergence of Newton's method in the Banach space X. For this, we proceed somewhat differently to the view in Section 2.1.1. In particular, we follow, from the general semilocal convergence result given in Theorem 1.27, the modification of the majorant principle of Kantorovich introduced Section 1.1.3 and construct a majorant function for the operator F. Next, we study the features of the majorant function and establish a semilocal convergence result for Newton's method under the following conditions:

(T1) There exists $\Gamma_0 = [F'(x_0)]^{-1} \in \mathcal{L}(Y, X)$, for some $x_0 \in \Omega$, with $\|\Gamma_0\| \leq \beta$ and $\|\Gamma_0 F(x_0)\| \leq \eta$.

(T2) There exists a continuous nondecreasing function $\varpi : [t_0, +\infty) \longrightarrow \mathbb{R}$ such that $\|F''(x)\| \leq \varpi(t; \|x_0\|, t_0)$, for $\|x - x_0\| \leq t - t_0$ and $\varpi(t_0; 0, t_0) = 0$.

2.2.1 Existence of a majorant function

For constructing a majorant function f for the operator F, in the sense of Definition 1.26 of Section 1.1.4, rely on Theorem 1.27 and the fact that the scalar function f must satisfy

$$\|F''(x)\| \leq f''(t), \quad \text{for} \quad \|x - x_0\| \leq t - t_0,$$

so that we choose $f''(t) = \varpi(t; \|x_0\|, t_0)$, for $\|x - x_0\| \leq t - t_0$, and then

$$\int_{t_0}^{t} f''(\xi)\, d\xi = \int_{t_0}^{t} \varpi(\xi; \|x_0\|, t_0)\, d\xi.$$

In addition, we can then obtain the function

$$f(t) = \int_{t_0}^{t} \int_{t_0}^{\theta} \varpi(\xi; \|x_0\|, t_0)\, d\xi\, d\theta - \frac{t - t_0}{\beta} + \frac{\eta}{\beta} \tag{2.32}$$

as the unique solution of the initial value problem

$$\begin{cases} y''(t) - \varpi(t; \|x_0\|, t_0) = 0, \\ y(t_0) = \dfrac{\eta}{\beta}, \quad y'(t_0) = -\dfrac{1}{\beta}. \end{cases} \tag{2.33}$$

Once seen how the scalar function must be to be majorant in the sense of Section 1.1.4 and under conditions (T1)-(T2), we establish, from the above-mentioned, the following result that proves the existence of such function.

Theorem 2.24. *We suppose that $\varpi(t; \|x_0\|, t_0)$ is continuous for all $t \in [t_0, +\infty)$. Then, for any nonnegative real numbers $\beta \neq 0$ and η, there exists only one solution $f(t)$ of initial value problem (2.33) in $[t_0, +\infty)$; that is, function (2.32), where ϖ is the function defined in (2.31).*

Note that (2.32) with $t_0 = 0$ is reduced to Kantorovich's polynomial (1.23) if ϖ is the constant M.

We have seen that Kantorovich constructs a majorizing real sequence $\{s_n\}$ from the application of Newton's method to the polynomial p defined in (1.23), so that $\{s_n\}$ converges to the smallest positive zero of the polynomial p. The convergence of sequence $\{s_n\}$ is obvious, since the polynomial p is a nonincreasing convex function in $[0, s^*]$, see Theorem 1.23.

Therefore, if we want to apply the method of majorizing sequences to our particular problem, the equation $f(t) = 0$, where f is defined in (2.32), must have at least one solution greater than t_0, so that we have to guarantee the convergence of the real sequence

$$t_{n+1} = N_f(t_n) = t_n - \frac{f(t_n)}{f'(t_n)}, \quad n = 0, 1, 2, \ldots, \tag{2.34}$$

from t_0, to this solution, for obtaining a majorizing sequence under conditions (T1)-(T2). Clearly, the first we need is to analyse the function f defined in (2.32). Then, we give some properties of this f.

Theorem 2.25. *Let f and ϖ be the functions defined respectively in (2.32) and (2.31).*

(a) *There exists only one positive solution $\alpha > t_0$ of the equation*

$$f'(t) = \int_{t_0}^{t} \varpi(\xi; \|x_0\|, t_0) \, d\xi - \frac{1}{\beta} = 0, \tag{2.35}$$

which is the unique minimum of $f(t)$ in $[t_0, +\infty)$ and $f(t)$ is nonincreasing in $[t_0, \alpha)$.

(b) *If $f(\alpha) \leq 0$, then the equation $f(t) = 0$ has at least one solution in $[t_0, +\infty)$. Moreover, if t^* is the smallest solution of $f(t) = 0$ in $[t_0, +\infty)$, we have $t_0 < t^* \leq \alpha$.*

Proof. First, as $f'(\alpha) = \int_{t_0}^{\alpha} \varpi(\xi; \|x_0\|, t_0) \, d\xi - \frac{1}{\beta} = 0$ and $f''(t) = \varpi(\xi; \|x_0\|, t_0) \geq 0$ in $[t_0, +\infty)$, then α is a minimum of f in $[t_0, +\infty)$. Moreover, as f is convex in $[t_0, +\infty)$, then α is the unique minimum of f in $[t_0, +\infty)$ and f' is nondecreasing in $[t_0, +\infty)$.

Second, as $f'(t_0) = -\frac{1}{\beta} < 0$, $f'(\alpha) = 0$ and f' is nondecreasing in $[t_0, +\infty)$, then $f'(t) < 0$ in $[t_0, \alpha)$, so that f is nonincreasing in $[t_0, \alpha)$.

Third, if $f(\alpha) < 0$, as $f(t_0) = \frac{\eta}{\beta} > 0$ and f is continuous, then f has at least one zero t^* in (t_0, α). As f is nonincreasing in $[t_0, \alpha)$, then t^* is the unique zero of f in (t_0, α).

Finally, if $f(\alpha) = 0$, then α is a double zero of f and $t^* = \alpha$. ∎

2.2.2 Existence of a majorizing sequence

As we are interested in the fact that (2.34) is a majorizing sequence of the sequence $\{x_n\}$ defined by Newton's method in the Banach space X, we establish the convergence of $\{t_n\}$ in the next result. Note that the function f given in (2.32) has the same properties as Kantorovich's polynomial (1.23), since $f(t) > 0$, $f'(t) < 0$ and $f''(t) > 0$ in (t_0, α), so that the convergence of the real sequence $\{t_n\}$ is then guaranteed from the next theorem, whose proof is completely analogous to that of Theorem 1.21.

Theorem 2.26. *Let $\{t_n\}$ be the real sequence defined in (2.34), where the function f is given in (2.32). Suppose that there exist a solution $\alpha > t_0$ of equation (2.35) such that $f(\alpha) \leq 0$. Then, the real sequence $\{t_n\}$ is nondecreasing and converges to the solution t^* of the equation $f(t) = 0$.*

The following is to prove that (2.34) is a majorizing sequence of the sequence $\{x_n\}$ and this sequence is well-defined, provided that $B(x_0, t^* - t_0) \subseteq \Omega$. Before, taking into account Theorem 2.26, we introduce the condition:

(T3) $f(\alpha) \leq 0$, where f is the function given in (2.32) and α is a solution of $f'(t) = 0$ such that $\alpha > t_0$, and $B(x_0, t^* - t_0) \subset \Omega$, where t^* is the smallest positive solution of $f(t) = 0$ in $[t_0, +\infty)$.

Theorem 2.27. *Let f be the function defined in (2.32). Suppose that conditions (T1)-(T2)-(T3) are satisfied. Then, $x_n \in B(x_0, t^* - t_0)$, for all $n \in \mathbb{N}$. Moreover, (2.34) is a majorizing sequence of the sequence $\{x_n\}$, namely*

$$\|x_n - x_{n-1}\| \leq t_n - t_{n-1}, \quad \text{for all} \quad n \in \mathbb{N}.$$

Proof. First of all, from (A1), we observe that

$$\|x_1 - x_0\| = \|\Gamma_0 F(x_0)\| \leq \eta = t_1 - t_0 < t^* - t_0.$$

After that, we prove the theorem from the next four recurrence relations (for $n \in \mathbb{N}$):

(i_n) There exists $\Gamma_n = [F'(x_n)]^{-1}$ and $\|\Gamma_n\| \leq -\dfrac{1}{f'(t_n)}$,

(ii_n) $\|F(x_n)\| \leq f(t_n)$,

(iii_n) $\|x_{n+1} - x_n\| \leq t_{n+1} - t_n$,

(iv_n) $\|x_{n+1} - x_0\| \leq t^* - t_0$.

We begin proving (i_1)-(ii_1)-$(iiii_1)$-(iv_1). First, from $x = x_0 + \tau(x_1 - x_0)$ and $t = t_0 + \tau(t_1 - t_0)$, where $0 \leq \tau \leq 1$, it follows $\|x - x_0\| = \tau\|x_1 - x_0\| \leq \tau(t_1 - t_0) = t - t_0$, so that

$$\begin{aligned}
\|I - \Gamma_0 F'(x_1)\| &= \left\| \int_{x_0}^{x_1} \Gamma_0 F''(x) \right\| \\
&= \left\| \int_0^1 \Gamma_0 F''(x_0 + t(x_1 - x_0))(x_1 - x_0)\, dt \right\| \\
&\leq \|\Gamma_0\| \int_0^1 \|F''(x_0 + \tau(x_1 - x_0))\|\, d\tau \|x_1 - x_0\| \\
&\leq \|\Gamma_0\| \int_0^1 \varpi(t_0 + \tau(t_1 - t_0); \|x_0\|, t_0)\, d\tau(t_1 - t_0) \\
&\leq \beta \int_{t_0}^{t_1} \varpi(t; \|x_0\|, t_0)\, dt \\
&= \beta \int_{t_0}^{t_1} f''(t)\, dt \\
&= 1 - \frac{f'(t_1)}{f'(t_0)} \\
&< 1,
\end{aligned}$$

since $\beta = -\frac{1}{f'(t_0)}$ and $\|x_1 - x_0\| \leq t_1 - t_0$. Then, from the Banach lemma on invertible operators, we obtain that there exists Γ_1 and $\|\Gamma_1\| \leq -\frac{1}{f'(t_1)}$.

Second, from Taylor's series, $\|x_1 - x_0\| \leq t_1 - t_0$ and the algorithm of Newton's method, we have

$$
F(x_1) = F(x_0) + F'(x_0)(x_1 - x_0) + \int_{x_0}^{x_1} F''(x)(x_1 - x)\,dx
$$
$$
= \int_{x_0}^{x_1} F''(x)(x - x_0)\,dx
$$
$$
= \int_0^1 F''(x_0 + \tau(x_1 - x_0))\,\tau\,d\tau\,(x_1 - x_0)^2
$$

and

$$
\|F(x_1)\| \leq \int_0^1 \varpi\,(t_0 + \tau(t_1 - t_0); \|x_0\|, t_0)\,\tau\,d\tau(t_1 - t_0)^2
$$
$$
= \int_0^1 f''\,(t_0 + \tau(t_1 - t_0))\,\tau\,d\tau(t_1 - t_0)^2
$$
$$
= \int_{t_0}^{t_1} f''(\tau)(t_1 - \tau)\,d\tau
$$
$$
= f(t_1) - f(t_0) - f'(t_0)(t_1 - t_0)
$$
$$
= f(t_1).
$$

Third, $\|x_2 - x_1\| \leq \|\Gamma_1\|\|F(x_1)\| \leq -\frac{f(t_1)}{f'(t_1)} = t_2 - t_1$.

Forth, $\|x_2 - x_0\| \leq \|x_2 - x_1\| + \|x_1 - x_0\| \leq t_2 - t_0 \leq t^* - t_0$.

After that, if we now assume that (i_j)-(ii_j)-(iii_j)-(iv_j) are true for $j = 1, 2, \ldots, m$, we can prove that (i_{m+1})-(ii_{m+1})-(iii_{m+1})-(iv_{m+1}) hold, so that (i_n)-(ii_n)-(iii_n)-(iv_n) are true for all positive integer n by mathematical induction. ∎

2.2.3 Semilocal convergence

We are then ready to prove the following semilocal convergence result for Newton's method under conditions (T1)-(T2)-(T3).

Theorem 2.28. *Let $F : \Omega \subseteq X \longrightarrow Y$ be a twice continuously Fréchet differentiable operator defined on a nonempty open convex domain Ω of a Banach space X with values in a Banach space Y. Suppose that conditions (T1)-(T2)-(T3) are satisfied. Then, Newton's sequence $\{x_n\}$ converges to a solution x^* of $F(x) = 0$ starting at x_0. Moreover, $x_n, x^* \in \overline{B(x_0, t^* - t_0)}$ and*

$$
\|x^* - x_n\| \leq t^* - t_n, \quad n = 0, 1, 2, \ldots
$$

Proof. Observe that $\{x_n\}$ is convergent, since $\{t_n\}$ is a majorizing real sequence of $\{x_n\}$ and convergent. Moreover, as $\lim_n t_n = t^*$, if $x^* = \lim_n x_n$, then $\|x^* - x_n\| \leq t^* - t_n$, for all

$n = 0, 1, 2, \ldots$ Furthermore,

$$
\begin{aligned}
\|F'(x_n) - F'(x_0)\| &= \left\| \int_{x_0}^{x_n} F''(x)\, dx \right\| \\
&= \left\| \int_0^1 F''(x_0 + \tau(x_n - x_0))\, d\tau(x_n - x_0) \right\| \\
&\le \int_0^1 \|F''(x_0 + \tau(x_n - x_0))\|\, d\tau \|x_n - x_0\| \\
&\le \int_0^1 \varpi\, (t_0 + \tau(t_n - t_0); \|x_0\|, t_0)\, d\tau(t^* - t_0) \\
&\le \int_0^1 \varpi\, (t_0 + \tau(t^* - t_0); \|x_0\|, t_0)\, d\tau(t^* - t_0) \\
&\le \int_0^1 \varpi\, (t^*; \|x_0\|, t_0)\, d\tau(t^* - t_0) \\
&\le \varpi(t^*; \|x_0\|, t_0)(t^* - t_0),
\end{aligned}
$$

since $\|x - x_0\| \le t - t_0$, $\|x_n - x_0\| \le t^* - t_0$ and $t_0 + \tau(t_n - t_0) < t^*$, for all $n = 0, 1, 2, \ldots$, so that

$$
\|F'(x_n)\| \le \|F'(x_0)\| + \|F'(x_n) - F'(x_0)\| \le \|F'(x_0)\| + \varpi(t^*; \|x_0\|, t_0)(t^* - t_0),
$$

and, consequently, the sequence $\{\|F'(x_n)\|\}$ is bounded. Therefore, from

$$
\|F(x_n)\| = \|F'(x_n)(x_{n+1} - x_n)\| \le \|F'(x_n)\|\|x_{n+1} - x_n\|,
$$

it follows that $\lim_n \|F(x_n)\| = 0$, and, by the continuity of F, we obtain $F(x^*) = 0$. ∎

2.2.4   Uniqueness of solution and order of convergence

Once we have proved the semilocal convergence of Newton's method and located the solution x^*, we prove the uniqueness of x^* and the quadratic convergence of Newton's method under conditions (T1)-(T2)-(T3). For this, we study function (2.32).

First, we note that if $\varpi(t_0; \|x_0\|, t_0) > 0$, then f' is nondecreasing and $f'(t) > 0$ in $(a, +\infty)$, since ϖ is nondecreasing and $\varpi(t; \|x_0\|, t_0) = f''(t) > 0$. Therefore, f is strictly nondecreasing and convex in $(a, +\infty)$. The last guarantees that f has two real zeros t^* and t^{**} such that $t_0 < t^* \le t^{**}$. Second, if $\varpi(t_0; \|x_0\|, t_0) = 0$ and $\varpi(t'; \|x_0\|, t_0) > 0$, for some $t' > t_0$, then it takes place the same as in the previous case. Third, finally, if $\varpi(t_0; \|x_0\|, t_0) = 0$ and $\varpi(t; \|x_0\|, t_0) = 0$, for all $t > t_0$, then f is lineal. Note that the latter is not restrictive because only the lineal case is eliminated. Observe that this case is trivial: if $f(t) = at + b$, then $N_f(t) = -\frac{b}{a}$, which is the solution of $f(t) = 0$.

After that, we establish the uniqueness of solution in the following result, whose proof is similar to that given in the Huang theorem, Theorem 2.3, without more that take into account (T2) with $\varpi(t; \|x_0\|, t_0) = f''(t)$ instead of (2.5).

Theorem 2.29. *Under the conditions of Theorem 2.28, the solution x^* is unique in $B(x_0, t^{**} - t_0) \cap \Omega$ if $t^* < t^{**}$ or in $\overline{B(x_0, t^* - t_0)}$ if $t^* = t^{**}$.*

Next, we see that the quadratic convergence of Newton's method is also guaranteed under conditions (T1)-(T2)-(T3). Note first that if the function f defined in (2.32) has two real positive zeros t^* and t^{**} such that $t^* \leq t^{**}$, we can then write

$$f(t) = (t^* - t)(t^{**} - t)g(t)$$

with $g(t^*) \neq 0$ and $g(t^{**}) \neq 0$.

Then, we give a result which provides some error estimates that lead to the quadratic convergence of Newton's method. The proof of the result is completely analogous to that of Theorem 2.18. Remember that we have written at the beginning of this section how the function ϖ should be for f to have two real positive solutions.

Theorem 2.30. *Under the hypotheses of Theorem 2.28 and assuming that the function f defined in (2.32) is such that $f(t) = (t^* - t)(t^{**} - t)g(t)$, where $g(t)$ is a function such that $g(t^*) \neq 0$ and $g(t^{**}) \neq 0$, we have the following:*

(a) *If $t^* < t^{**}$ and there exist $m_1 > 0$ and $M_1 > 0$ such that $m_1 \leq \min\{Q_1(t); t \in [t_0, t^*]\}$ and $M_1 \geq \max\{Q_1(t); t \in [t_0, t^*]\}$, where $Q_1(t) = \frac{(t^{**}-t)g'(t)-g(t)}{(t^*-t)g'(t)-g(t)}$, then*

$$\frac{(t^{**} - t^*)\theta^{2^n}}{m_1 - \theta^{2^n}} \leq t^* - t_n \leq \frac{(t^{**} - t^*)\Delta^{2^n}}{M_1 - \Delta^{2^n}}, \quad n \geq 0,$$

where $\theta = \frac{t^*}{t^{**}}m_1$, $\Delta = \frac{t^*}{t^{**}}M_1$ and provided that $\theta < 1$ and $\Delta < 1$.

(b) *If $t^* = t^{**}$ and there exist $m_2 > 0$ and $M_2 > 0$ such that $m_2 \leq \min\{Q_2(t); t \in [t_0, t^*]\}$ and $M_2 \geq \max\{Q_2(t); t \in [t_0, t^*]\}$, where $Q_2(t) = \frac{(t^*-t)g'(t)-g(t)}{(t^*-t)g'(t)-2g(t)}$, then*

$$m_2^n t^* \leq t^* - t_n \leq M_2^n t^*, \quad n \geq 0,$$

provided that $m_2 < 1$ and $M_2 < 1$.

From the last theorem, it follows that the convergence of Newton's method, under conditions (T1)-(T2), is quadratic if $t^* < t^{**}$ and linear if $t^* = t^{**}$.

Remark 2.31. Finally, we observe that condition (U2),

$$\|F''(x) - F''(x_0)\| \leq \omega_0(\|x - x_0\|), \quad x \in \Omega,$$

involves condition (2.30),

$$\|F''(x)\| \leq \omega(\|x\|), \quad x \in \Omega,$$

since

$$\|F''(x)\| \leq \omega_0(\|x - x_0\|) + \|F''(x_0)\| \leq \omega_0(\|x\| + \|x_0\|) + \|F''(x_0)\|$$

and we can then choose $w(z) = w_0(z + \|x_0\|) + \|F''(x_0)\|$.

2.2.5 Applications

In this section, we present two applications where we can see some advantages of the previous study of Newton's method with respect to that made by Kantorovich from the Newton-Kantorovich theorem. In the first example, we present how the domain of starting points, the domains of existence and uniqueness of solution and the a priori error bounds are improved. In the second example, we see that the Newton-Kantorovich theorem cannot be applied to approximate a solution of integral equation (1.40) because the solution of the equation cannot be located previously, while we can do it with Theorem 2.28.

Notice that, in practice, we usually choose $t_0 = 0$, since function (2.32) has the same property of translation as Kantorovich's polynomial (see Remark 1.18) and majorizes the operator F, in the sense of Definition 1.26, independent of the value of t_0.

Example 2.32. By means of the following simple example, we show that we can improve the domain of starting points, the domains of existence and uniqueness of solution and the error bounds for Newton's method if we use condition (T2) instead of condition (A2).

Consider the equation $F(x) = 0$, where $F : \Omega = (0, a) \longrightarrow \mathbb{R}$ and $F(x) = x^3 - a$ with $a > 1$. Then,

$$\|\Gamma_0\| = \frac{1}{3x_0^2} = \beta, \quad \|\Gamma_0 F(x_0)\| = \frac{|x_0^3 - a|}{3x_0^2} = \eta, \quad \|F''(x)\| = 6|x|.$$

As a consequence, we have $M = 6a$ for the Newton-Kantorovich theorem and $\varpi(t; \|x_0\|, t_0) = \varpi(t; \|x_0\|, 0) = \omega(t + \|x_0\|) = 6(t + |x_0|)$ for Theorem 2.28.

When analyzing the domain of starting points for Newton's method from the Newton-Kantorovich theorem and Theorem 2.28, we will only pay attention to the interval $(0, \sqrt[3]{a})$, since Newton's method always converges if we choose x_0 in the interval $(\sqrt[3]{a}, a)$, since F is nondecreasing and convex in $(\sqrt[3]{a}, a)$, see Figure 2.6.

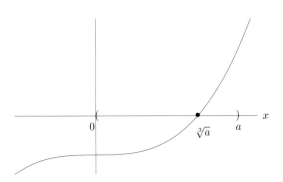

Figure 2.6: $F(x) = x^3 - a$

For the Newton-Kantorovich theorem, we need that $M\beta\eta \le \frac{1}{2}$, which is equivalent to $3x_0^4 + 4ax_0^3 - 4a^2 \equiv g(x_0) \ge 0$, since $x_0 \in (0, \sqrt[3]{a})$. In addition, $x_0 \in (r^*, \sqrt[3]{a})$, where r^* is such that $g(r^*) = 0$. For Theorem 2.28, we need that

$$0 \ge f(a) = (5 - 4\sqrt{2})|x_0|^3 + |x_0^3 - a|,$$

where $f(t) = (t + |x_0|)^3 - 6|x_0|^2 t - |x_0|^3 + |x_0^3 - a|$, with $t_0 = 0$, and $\alpha = (\sqrt{2} - 1)|x_0|$, which is equivalent to $4(1 - \sqrt{2})x_0^3 + a \leq 0$, since $x_0 \in (0, \sqrt[3]{a})$. Therefore, $x_0 \geq \sqrt[3]{\frac{a}{4(\sqrt{2}-1)}}$.

If we consider the particular case $a = 2011$, we obtain $x_0 \in (12.6026\ldots, \sqrt[3]{2011})$ for the Newton-Kantorovich theorem and $x_0 \in (10.6670\ldots, \sqrt[3]{2011})$ for Theorem 2.28. Therefore, we improve the domain of starting points by Theorem 2.28 with respect to the Newton-Kantorovich theorem.

Taking then, for example, $x_0 = 12.61$, we obtain $s^* = 0.01520\ldots$ and $s^{**} = 0.06387\ldots$ for the Newton-Kantorovich theorem, so that the domains of existence and uniqueness of solution are respectively

$$\{z \in (0, 2011) : |z - x_0| \leq 0.01520\ldots\} \quad \text{and} \quad \{z \in (0, 2011) : |z - x_0| < 0.06387\ldots\}.$$

For Theorems 2.28 and 2.29, we have that $t^* = 0.01229\ldots$ and $t^{**} = 9.96796\ldots$ are the two zeros of the majorant function $f(t)$ with $t_0 = 0$, so that the domains of existence and uniqueness of solution are respectively

$$\{z \in (0, 2011) : |z - x_0| \leq 0.01229\ldots\} \quad \text{and} \quad \{z \in (0, 2011) : |z - x_0| < 9.96796\ldots\}.$$

Hence, the domains of existence and uniqueness of solution that we have obtained from Theorems 2.28 and 2.29 are better than those obtained from the Newton-Kantorovich theorem.

We also obtain better error bounds from majorizing real sequences (see Table 2.14), where $\{s_n\}$ denotes the majorizing real sequence obtained from Kantorovich's polynomial (1.23) and $\{t_n\}$ denotes the majorizing real sequence obtained from the majorant function $f(t)$.

n	$|x^* - x_n|$	$|t^* - t_n|$	$|s^* - s_n|$
0	$1.2266\ldots \times 10^{-2}$	$1.2290\ldots \times 10^{-2}$	$1.5201\ldots \times 10^{-2}$
1	$1.1936\ldots \times 10^{-5}$	$1.1983\ldots \times 10^{-5}$	$2.9223\ldots \times 10^{-3}$
2	$1.1290\ldots \times 10^{-11}$	$1.1421\ldots \times 10^{-11}$	$1.5666\ldots \times 10^{-4}$

Table 2.14: Absolute error and error bounds for $x^* = \sqrt[3]{2011}$

Example 2.33. In this example, we use what we have seen so far in Section 2.2 to guarantee the convergence of Newton's method to a solution $x^{**}(s)$ of integral equation (1.40),

$$x(s) = s^2 + s \int_0^1 t^{13} \left(x(t)^{7/3} + \frac{1}{2}x(t)^2 \right) dt, \quad s \in [0, 1],$$

such that $\|x^{**}(s)\| \geq \rho_2 = 5.0672\ldots$ We already know that the Newton-Kantorovich theorem cannot guarantee this convergence, since a domain where $x^{**}(s)$ lies and $\|\mathcal{F}''(x)\|$ is bounded cannot be fixed. But Theorem 2.28 can do it, as we see below. Notice first that in this case $\Omega = \mathcal{C}([0, 1])$.

In view of how it is the solution $x^*(s)$ of equation (1.40) and taking into account that $\|x^{**}(s)\| \geq \rho_2 = 5.0672\ldots$, a reasonable choice of the initial approximation seems to be $x_0(s) = 5s + s^2$, that satisfies $\|x_0(s)\| = 6 > \rho_2 = 5.0672\ldots$

For Theorem 2.28, we have $\|\mathcal{F}''(x)\| \leq \omega(\|x\|)$, where $\omega(t) = \frac{2}{9}\sqrt{t+6} + \frac{1}{14}$, so that

$$f(t) = (0.0714\ldots)\left(-(65.1439\ldots) - (29.7659\ldots)t + \frac{t^2}{2} + (36 + 12t + t^2)\sqrt[3]{t+6} \right),$$

with $t_0 = 0$, and $t^* = 0.0663\ldots$ and $t^{**} = 1.2105\ldots$ are the two positive real solutions. The domains of existence and uniqueness of solution are, respectively, in this case

$$\{\nu \in \Omega : \|\nu(s) - x_0(s)\| \leq 0.0663\ldots\} \quad \text{and} \quad \{\nu \in \Omega : \|\nu(s) - x_0(s)\| < 1.2105\ldots\}.$$

Note that condition (1.42) is satisfied in $B(x_0, t^*)$ for all $x(s)$, so that the sequence $\{x_n(s)\}$, given by Newton's method, is well-defined.

After three iterations of Newton's method, we obtain the approximated solution $x^{**}(s) = (4.9784\ldots)s + s^2$ given in Table 2.15. From the sequence $\{\|[\mathcal{F}(x_n)](s)\|\}$ given in Table 2.15, we observe that $x^{**}(s)$ is a good approximation of a solution of (1.40). In Table 2.16, we show the errors $\|x^{**}(s) - x_n(s)\|$ and the error bounds. Moreover, observe that $\|x^{**}(s)\| = 5.9784\ldots > \rho_2 = 5.0672\ldots$, so that $x^{**}(s)$ is a solution that is beyond the scope of the Newton-Kantorovich theorem.

n	$x_n(s)$	$\|[\mathcal{F}(x_n)](s)\|$
0	$5s + s^2$	$1.9455\ldots \times 10^{-2}$
1	$(4.978595065161226\ldots)s + s^2$	$9.3321\ldots \times 10^{-5}$
2	$(4.978491401412390\ldots)s + s^2$	$2.1873\ldots \times 10^{-9}$
3	$(4.978491398982550\ldots)s + s^2$	

Table 2.15: Approximated solution $x^{**}(s)$ of (1.40) and $\{\|[\mathcal{F}(x_n)](s)\|\}$

n	$\|x^{**}(s) - x_n(s)\|$	$|t^* - t_n|$
0	$2.1508\ldots \times 10^{-2}$	$6.7508\ldots \times 10^{-2}$
1	$1.0366\ldots \times 10^{-4}$	$4.5480\ldots \times 10^{-3}$
2	$2.4298\ldots \times 10^{-7}$	$2.3684\ldots \times 10^{-5}$

Table 2.16: Absolute errors and error bounds

Moreover, we observe that we can also use Theorem 2.28 to obtain better domains of existence and uniqueness of solution and better error bounds than those obtained from the Newton-Kantorovich theorem when Newton's method approximates the first solution $x^*(s)$ of integral equation (1.40) which is given in Table 1.1.

If we choose $t_0 = 0$ for Theorem 2.28, then

$$f(t) = (0.0714\ldots)\left((0.1388\ldots) - (13.5308\ldots)t + \frac{t^2}{2} + (1 + 2t + t^2)\sqrt[3]{t+1}\right)$$

and $t^* = 0.1037\ldots$ and $t^{**} = 4.2856\ldots$ are the two positive real solutions. As a consequence, the domains of existence and uniqueness of solution are respectively

$$\left\{u \in \mathcal{C}([0,1]) : \|u(s) - s^2\| \leq 0.1037\ldots\right\} \quad \text{and} \quad \left\{u \in \mathcal{C}([0,1]) : \|u(s) - s^2\| < 4.2856\ldots\right\}.$$

Notice that the domains obtained from Theorem 2.28 and 2.29 improve those given by the Newton-Kantorovich theorem (see Section 1.2), since the domain of existence of solution is smaller and the domain of uniqueness of solution is bigger.

n	$\|x^*(s) - x_n(s)\|$	$|t^* - t_n|$	$|s^* - s_n|$
0	$1.0216\ldots \times 10^{-1}$	$1.0370\ldots \times 10^{-1}$	$1.0437\ldots \times 10^{-1}$
1	$1.6214\ldots \times 10^{-3}$	$1.9910\ldots \times 10^{-3}$	$2.6693\ldots \times 10^{-3}$
2	$1.8373\ldots \times 10^{-7}$	$7.7520\ldots \times 10^{-7}$	$1.8373\ldots \times 10^{-6}$
3	$3.0004\ldots \times 10^{-14}$	$1.1764\ldots \times 10^{-13}$	$8.7172\ldots \times 10^{-13}$

Table 2.17: Absolute errors and error bounds

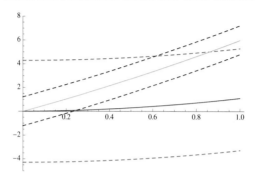

Figure 2.7: Graphs of the approximated solutions $x^*(s)$ and $x^{**}(s)$ of (1.40) (respectively, blue solid line and green solid line), along with their domains of uniqueness (respectively, red dashed lines and black dashed lines)

In Table 2.17, we can see that the error bounds $\{|t^* - t_n|\}$ obtained from Theorem 2.28 improves those, $\{|s^* - s_n|\}$, given by the majorizing real sequence obtained from Kantorovich's polynomial (1.23).

Finally, we see in Figure 2.7 the approximated solutions $x^*(s)$ and $x^{**}(s)$ of (1.40) along with the uniqueness domains of solution obtained above. Observe that the uniqueness domains separate perfectly the two solutions.

On the other hand, from Theorem 2.30, analogous conclusions to those of Example 2.20 about the a priori error bounds can be drawn.

Chapter 3

Convergence conditions on the k-th derivative of the operator

Following with the idea of improving the applicability of Newton's method, in this chapter, we extend the semilocal convergence results seen in the previous chapter for Newton's method under conditions on the second derivative of the operator involved. So, we establish semilocal convergence results for Newton's method under conditions on derivatives of the operator of order greater than two.

We start generalizing condition (A2) of the Newton-Kantorovich theorem with the condition $\|F^{(k)}(x)\| \leq \ell$, for $x \in \Omega$ and $k \geq 3$ [27]. Note that, if we consider polynomial operators, although $F''(x)$ is not bounded, there is always a value of k such that $F^{(k)}(x) = \text{constant}$, so that we will be able to consider the previous condition for $F^{(k)}(x)$.

After that, we generalize the last condition for $F^{(k)}(x)$ to the condition $\|F^{(k)}(x)\| \leq \omega(\|x\|)$, $x \in \Omega$, where $\omega : [0, +\infty) \longrightarrow \mathbb{R}$ is a continuous monotonous function such that $\omega(0) \geq 0$ [28].

Next, we require a more general condition on $F^{(k)}(x)$: $F^{(k)}(x)$ is ω-Lipschitz continuous in Ω. We have already seen in Chapter 2 that a condition of ω-Lipschitz continuity implies a condition of ω-boundedness. In addition, we finish the chapter requiring a center ω-Lipschitz condition to the operator $F^{(k)}(x)$ [29], a condition that arises prior to studying the case in which $F^{(k)}(x)$ is ω-Lipschitz continuous in Ω.

3.1   Polynomial type equations

If we consider the nonlinear conservative system given by (1.44)-(1.45),

$$\frac{d^2x(s)}{ds^2} + \phi(x(s)) = 0, \qquad x(0) = x(1) = 0,$$

we have seen in Section 1.2.2 that we can then consider the second-order differential equation as a twice differentiable operator $\mathfrak{F} : \mathcal{C}^2([0, 1]) \longrightarrow \mathcal{C}([0, 1])$ such that

$$[\mathfrak{F}(x)](s) = \frac{d^2x(s)}{ds^2} + \phi(x(s))$$

and provided that $\phi \in \mathcal{C}^2(\mathbb{R})$. The first two derivatives of the operator $\mathfrak{F}$ are

$$\mathfrak{F}'(x) = \frac{d^2}{ds^2} + \phi'(x(s))I, \qquad \mathfrak{F}''(x) = \phi''(x(s))I_2,$$

© Springer International Publishing AG 2017
J.A. Ezquerro Fernández, M.Á. Hernández Verón, *Newton's Method: an Updated Approach of Kantorovich's Theory*, Frontiers in Mathematics, DOI 10.1007/978-3-319-55976-6_3

where I_2 is the bilinear operator defined by $I_2 xy(s) = x(s)y(s)$.

Observe that condition (A2) of Kantorovich is not satisfied in general because $\|\mathfrak{F}''(x)\|$ is not bounded in $\mathcal{C}^2([0,1])$ unless $\phi(u)$ is a polynomial function of degree less than or equal to two. In other case, it is not easy to find a domain $\Omega \subseteq \mathcal{C}^2([0,1])$ where $\|\mathfrak{F}''(x)\|$ is bounded and contains a solution of the equation to solve.

To solve the last problem, we known that a common alternative is to locate previously a solution $x^*(s)$ of problem (1.44)-(1.45) in a domain $\Omega \subseteq \mathcal{C}^2([0,1])$ and look for a bound for $\|\mathfrak{F}''(x)\|$ in Ω (see [23]). So, taking into account that a solution $x^*(s)$ of (1.44)-(1.45) is of the form defined in (1.46), this solution must satisfy

$$\|x^*(s)\| - S\|\phi(x^*(s))\| \leq 0, \tag{3.1}$$

where $S = \left\| \int_0^1 G(s,t)\, dt \right\|$. Therefore, from (3.1), we try to find a region $\Omega \subseteq \mathcal{C}^2([0,1])$ which contains $x^*(s)$. However, equation (3.1), that $x^*(s)$ must verify, can be satisfied in different regions of the space. For example, if the scalar equation deduced from (3.1) and given by $\chi(t) = t - S\|\phi(t)\| = 0$ has three positive roots r_1, r_2 and r_3, such that $r_1 < r_2 < r_3$, we can only locate previously the solutions $x^*(s)$ such that $\|x^*(s)\| \in [0, r_1]$ or $\|x^*(s)\| \in [r_2, r_3]$. In other case, namely, when $\|x^*(s)\| > r_3$, the problem remains open. Observe that the case in which the equation $\chi(t) = 0$ does not have positive real roots is analogous to the previous one. Other situations for the number of roots are reduced to some of the last ones.

In this section, we propose an alternative to locate previously the solutions of problem (1.44)-(1.45) and also respond to situations that the previous location of solutions does not solve. Notice that, as we have indicated, if $\phi(u) = a_0 + a_1 u + \cdots + a_k u^k$, $a_k \in \mathbb{R}$, we can apply the conditions given by Kantorovich in the case where $k \leq 2$, but not for $k > 2$. So, we generalize the hypotheses of Kantorovich by modifying conditions (A1)-(A2) in the following way:

(J1) There exists the operator $\Gamma_0 = [F'(x_0)]^{-1} \in \mathcal{L}(Y, X)$, for some $x_0 \in \Omega$, with $\|\Gamma_0\| \leq \beta$ and $\|\Gamma_0 F(x_0)\| \leq \eta$; moreover, $\|F^{(i)}(x_0)\| \leq b_i$, with $b_i \in \mathbb{R}_+$, $i = 2, 3, \ldots, k-1$ and $k \geq 3$.

(J2) There exists a constant $\ell \geq 0$ such that $\|F^{(k)}(x)\| \leq \ell$, for $x \in \Omega$.

Obviously, the above generalization of Kantorovich's conditions (A1)-(A2) leads to a variation in the method of majorizing sequences that is developed in Section 1.1.3 to prove the semilocal convergence of Newton's method under conditions (A1)-(A2). Specifically, Kantorovich's polynomial (1.23), $p(s) = \frac{M}{2} s^2 - \frac{s}{\beta} + \frac{\eta}{\beta}$, that Kantorovich obtains from conditions (A1)-(A2), for constructing a majorizing real sequence, is replaced by a polynomial of degree k, and condition (A3), $h = M\beta\eta \leq \frac{1}{2}$, which guarantees that the polynomial p has real positive roots, by the corresponding condition that guarantees that the polynomial of degree k has also real positive roots.

In [3], we can see an analogous study for polynomial equations in Banach spaces and based on ideas given in [2] and [37], but taking into account that a different technique to that of Kantorovich was used to prove the semilocal convergence of Newton's method. Here, we emphasize that we consider any operator in Banach spaces and follow Kantorovich's technique to prove the semilocal convergence of Newton's method. For this, we construct a majorant polynomial from a problem of interpolation fitting, as Kantorovich did for Kantorovich's polynomial (1.23). However, in Section 3.2, we study the semilocal convergence of Newton's method under a generalization of condition (J2) and, in this case, we are forced to use the modification of the majorant principle of Kantorovich presented previously and thus obtain the majorant function as a solution of an initial value problem. In addition, the situation presented in Section 3.1 is a particular case of that presented in Section 3.2.

The rest of Section 3.1 is divided in three subsections. In Section 3.1.1 we establish the semilocal convergence of Newton's method in Banach spaces under conditions (J1)-(J2). For this, we construct a majorizing real sequence which fits in with conditions (J1)-(J2). In Section 3.1.2, we also draw conclusions about the existence and uniqueness of solution from the theoretical significance of Newton's method and give a priori error estimates that lead to the quadratic convergence of Newton's method. We finish with Section 3.1.3, where two particular conservative problems are studied, emphasizing the disadvantage of Kantorovich's conditions with regard to the conditions introduced in Section 3.1. In advance, we can say that Kantorovich needs to locate previously the solutions of the equations and chooses good starting points for applying Newton's method, while we only need to choose good starting points to obtain the convergence of Newton's method, provided that the new conditions are considered.

3.1.1   Semilocal convergence

In this section we prove the semilocal convergence of Newton's method under conditions (J1)-(J2) and following the method of majorizing sequences developed in Section 1.1.3 for conditions (A1)-(A2). For this, we first construct a majorizing real sequence $\{t_n\}$ of Newton's sequence $\{x_n\}$ in the Banach space X.

To construct the majorizing real sequence $\{t_n\}$, we need to find a real function $f : I \subseteq \mathbb{R} \longrightarrow \mathbb{R}$ and consider that the real sequence

$$t_0 \in I, \qquad t_n = N_f(t_{n-1}) = t_{n-1} - \frac{f(t_{n-1})}{f'(t_{n-1})}, \qquad n \in \mathbb{N}, \tag{3.2}$$

is nondecreasing and convergent to a solution of the equation $f(t) = 0$.

Then, the first questions are: how the function f should be so that sequence (4.1) majorizes Newton's sequence in X and how to determine the interval I. We suppose that Kantorovich followed the reasoning given in the proof of Theorem 1.16, where quadratic polynomial (1.21) is found by solving a problem of interpolation fitting, and consider $I = [t_0, +\infty)$. Now, taking into account that the value t_0 can be translated to zero (Remark 1.18) and the imposition that the equation $f(t) = 0$ has a solution in $[t_0, +\infty)$ lead to condition (A3). The function f is reduced to the quadratic polynomial given in (1.23). So, we can use this polynomial to construct majorizing sequence (4.1) and obtain the semilocal convergence result established in the Newton-Kantorovich theorem for Newton's method.

The case analyzed in this section is analogous, but considering the interpolation problem defined by conditions (J1)-(J2) given for the operator F, so that we look for a real function f such that

$$f^{(k)}(t) = \ell, \quad -\frac{1}{f'(t_0)} = \beta, \quad -\frac{f(t_0)}{f'(t_0)} = \eta, \quad f^{(i)}(t_0) = b_i, \quad \text{with} \quad i = 2, \ldots, k-1.$$

As a consequence, we obtain that $f(t)$ is the following polynomial of degree k:

$$f(t) = \frac{\ell}{k!}(t - t_0)^k + \frac{b_{k-1}}{(k-1)!}(t - t_0)^{k-1} + \cdots + \frac{b_2}{2!}(t - t_0)^2 - \frac{t - t_0}{\beta} + \frac{\eta}{\beta}. \tag{3.3}$$

Obviously, polynomial (3.3) verifies conditions (J1)-(J2), which are the conditions of a Taylor's interpolation problem.

Notice that if F is such that $\|F^{(k)}(x)\| = $ constant, we can consider condition (J2) with k replaced by $k + 1$, so that $\ell = 0$ and polynomial (3.3) would be the same as that obtained from condition (J2) with k.

To construct the majorizing real sequence $\{t_n\}$ defined in (4.1), where $f(t)$ is polynomial (3.3), we know that it is necessary that (3.3) has at least one positive zero t^* such that $t^* \geq t_0$ and sequence (4.1) is nondecreasing and convergent to t^*. Following with the method of majorizing sequences, the convergence of Newton's sequence in the Banach space X is then guaranteed from the convergence of majorizing sequence (4.1). First, we begin studying polynomial (3.3) in the following result. From now on, we denote α as the minimum of polynomial (3.3) in $(t_0, +\infty)$. Note that there exists only one zero α of $f'(t)$ in $(t_0, +\infty)$, since $f'(t_0) = -\frac{1}{\beta} < 0$, $f''(t) > 0$ and $f'(t) > 0$ as $t \to +\infty$.

Theorem 3.1. *If $f(\alpha) \leq 0$, then the polynomial $f(t)$, given in (3.3), has two real zeros t^* and t^{**} such that $t_0 \leq t^* \leq \alpha \leq t^{**}$.*

Proof. Since $f'(t_0) = -\dfrac{1}{\beta} < 0$, $f''(t) > 0$ and $f'(t) > 0$ when $t \to +\infty$, there exists a unique solution, denoted by α, of $f'(t) = 0$ in $(t_0, +\infty)$ and, as a consequence, the equation $f(t) = 0$ has two real solutions t^* and t^{**} in $[t_0, +\infty)$ such that $t^* \leq \alpha \leq t^{**}$. ∎

Second, we establish that sequence (4.1) is nondecreasing and converges to t^*. Observe that the function f given in (3.3) has the same properties as Kantorovich's polynomial (1.23), since $f(t) > 0$, $f'(t) < 0$ and $f''(t) > 0$ in (t_0, α), so that the convergence of the real sequence $\{t_n\}$ is then guaranteed from the next theorem, whose proof is completely analogous to that of Theorem 1.21.

Theorem 3.2. *If $f(\alpha) \leq 0$, then sequence (4.1) is nondecreasing and converges to the root t^* of the equation $f(t) = 0$.*

Third, we prove a system of recurrence relations in the next theorem that guarantees that (4.1) is a majorizing real sequence of Newton's method.

Lemma 3.3. *Suppose that conditions (J1)-(J2) are satisfied, $x_n \in \Omega$, for all $n \geq 0$, and $f(\alpha) \leq 0$, where $f(t)$ is polynomial (3.3) and α the minimum of the polynomial in $(t_0, +\infty)$. Then, for all $n \in \mathbb{N}$, the next four items hold:*

(i_n) *There exists $\Gamma_n = [F'(x_n)]^{-1}$ and $\|\Gamma_n\| \leq -\dfrac{1}{f'(t_n)}$,*

(ii_n) $\|F^{(i)}(x_n)\| \leq f^{(i)}(t_n)$, *for $i = 2, 3, \ldots, k - 1$,*

(iii_n) $\|F(x_n)\| \leq f(t_n)$,

(iv_n) $\|x_{n+1} - x_n\| \leq t_{n+1} - t_n$.

Proof. We use mathematical induction on n to prove items (i_n)-(ii_n)-(iii_n)-(iv_n). To prove (i_1), we need first that $\Gamma_1 = [F'(x_1)]^{-1}$ exists. Indeed, from

$$
\|I - \Gamma_0 F'(x_1)\| = \left\| I - \Gamma_0 \left(\sum_{j=1}^{k-1} \frac{1}{(j-1)!} F^{(j)}(x_0)(x_1 - x_0)^{j-1} \right. \right.
$$
$$
\left. \left. + \frac{1}{(k-2)!} \int_{x_0}^{x_1} F^{(k)}(z)(x_1 - z)^{k-2}\, dz \right) \right\|
$$
$$
\leq \|\Gamma_0\| \left(\sum_{j=2}^{k-1} \frac{1}{(j-1)!} \|F^{(j)}(x_0)\| \|x_1 - x_0\|^{j-1} \right.
$$
$$
\left. + \frac{1}{(k-2)!} \int_0^1 \left\| F^{(k)}\left(x_0 + \tau(x_1 - x_0)\right) \right\| (1-\tau)^{k-2} d\tau\, \|x_1 - x_0\|^{k-1} \right)
$$
$$
\leq -\frac{1}{f'(t_0)} \left(\sum_{i=2}^{k-1} \frac{1}{(j-1)!} f^{(j)}(t_0)(t_1 - t_0)^{j-1} \right.
$$
$$
\left. + \frac{1}{(k-2)!} \int_0^1 f^{(k)}\left(t_0 + \tau(t_1 - t_0)\right)(1-\tau)^{k-2} d\tau\, (t_1 - t_0)^{k-1} \right)
$$
$$
= -\frac{1}{f'(t_0)} \left(\sum_{j=2}^{k-1} \frac{1}{(j-1)!} f^{(j)}(t_0)(t_1 - t_0)^{j-1} \right.
$$
$$
\left. + \frac{1}{(k-2)!} \int_{t_0}^{t_1} f^{(k)}(\xi)(t_1 - \xi)^{k-2} d\xi \right)
$$
$$
\leq -\frac{1}{f'(t_0)} \left(f'(t_1) - f'(t_0) \right)
$$
$$
= 1 - \frac{f'(t_1)}{f'(t_0)}
$$
$$
< 1,
$$

since $\frac{f'(t_1)}{f'(t_0)} \in (0,1)$, and the Banach lemma on invertible operators, it follows that there exists the operator Γ_1 and $\|\Gamma_1\| \leq -\frac{1}{f'(t_1)}$.

To prove (ii_1), we consider any $i \in \{2, 3, \ldots, k-1\}$ and Taylor's series, to see

$$\|F^{(i)}(x_1)\| = \left\|\sum_{j=1}^{k-i} \frac{1}{(j-1)!} F^{(i+j-1)}(x_0)(x_1 - x_0)^{j-1}\right.$$
$$\left. + \frac{1}{(k-1-i)!} \int_{x_0}^{x_1} F^{(k)}(z)(x_1 - z)^{k-1-i} \, dz\right\|$$
$$\leq \sum_{j=1}^{k-i} \frac{1}{(j-1)!} \|F^{(j+i-1)}(x_0)\| \|x_1 - x_0\|^{j-1}$$
$$+ \frac{1}{(k-1-i)!} \int_0^1 \|F^{(k)}(x_0 + \tau(x_1 - x_0))\| (1-\tau)^{k-2} d\tau \|x_1 - x_0\|^{k-i}$$
$$\leq \sum_{j=1}^{k-i} \frac{1}{(j-1)!} f^{(j+i-1)}(t_0)(t_1 - t_0)^{j-1}$$
$$+ \frac{1}{(k-1-i)!} \int_0^1 f^{(k)}(t_0 + \tau(t_1 - t_0))(1-\tau)^{k-2} d\tau (t_1 - t_0)^{k-i}$$
$$= \sum_{j=1}^{k-i} \frac{1}{(j-1)!} f^{(i+j-1)}(t_0)(t_1 - t_0)^{j-1}$$
$$+ \frac{1}{(k-1-i)!} \int_{t_0}^{t_1} f^{(k)}(\xi)(t_1 - \xi)^{k-i-1} \, d\xi$$
$$= f^{(i)}(t_1).$$

To prove (iii_1), we consider Newton's sequence and (4.1), and take into account Taylor's series:

$$\|F(x_1)\| = \left\|\sum_{j=0}^{k-1} \frac{1}{j} F^{(i)}(x_0)(x_1 - x_0)^j + \frac{1}{(k-1)!} \int_{x_0}^{x_1} F^{(k)}(z)(x_1 - z)^{k-1} \, dz\right\|$$
$$\leq \sum_{j=2}^{k-1} \frac{1}{j!} \|F^{(j)}(x_0)\| \|x_1 - x_0\|^j$$
$$+ \frac{1}{(k-1)!} \int_0^1 \|F^{(k)}(x_0 + \tau(x_1 - x_0))\| (1-\tau)^{k-1} d\tau \|x_1 - x_0\|^k$$
$$\leq \sum_{j=2}^{k-1} \frac{1}{j!} f^{(j)}(t_0)(t_1 - t_0)^j$$
$$+ \frac{1}{(k-1)!} \int_0^1 f^{(k)}(t_0 + \tau(t_1 - t_0))(1-\tau)^{k-1} d\tau (t_1 - t_0)^k$$
$$= \sum_{j=0}^{k-1} \frac{1}{j!} f^{(j)}(t_0)(t_1 - t_0)^j + \frac{1}{(k-1)!} \int_{t_0}^{t_1} f^{(k)}(\xi)(t_1 - \xi)^{k-1} \, d\xi$$
$$= f(t_1).$$

Item (iv_1) follows immediately, since

$$\|x_2 - x_1\| \leq \|\Gamma_1\| \|F(x_1)\| \leq -\frac{f(t_1)}{f'(t_1)} = t_2 - t_1.$$

If we now assume that (i_n)-(ii_n)-(iii_n)-(iv_n) are satisfied for $n = 1, 2, \ldots, j$, it follows in the same way that (i_{j+1})-(ii_{j+1})-(iii_{j+1})-(iv_{j+1}) are true, so that (i_n)-(ii_n)-(iii_n)-(iv_n) are satisfied for all positive integers n by mathematical induction. ∎

Note that (i_0), (ii_0) and (iv_0) of the previous lemma are obvious and (iii_0) is not necessary to prove (iv_0), since it follows from the initial condition $\|\Gamma_0 F(x_0)\| \leq \eta$.

Once we have proved that (4.1) is a majorizing real sequence of Newton's sequence in the Banach space X, we are then ready to guarantee the semilocal convergence of Newton's method. For this, we consider:

(J3) $f(\alpha) \leq 0$, where $f(t)$ is polynomial (3.3) and α is the unique positive solution of $f'(t) = 0$ such that $\alpha > t_0$, and $B(x_0, t^* - t_0) \subset \Omega$, where t^* is the smallest positive solution of $f(t) = 0$ in $[t_0, +\infty)$.

Theorem 3.4. *Let $F : \Omega \subseteq X \longrightarrow Y$ be a k $(k \geq 3)$ times continuously Fréchet differentiable operator defined on a non-empty open convex domain Ω of a Banach space X with values in a Banach space Y. Suppose that conditions (J1)-(J2)-(J3) are satisfied. Then Newton's sequence $\{x_n\}$ converges to a solution x^* of the equation $F(x) = 0$ and $x_n, x^* \in \overline{B(x_0, t^* - t_0)}$, for all $n = 0, 1, 2, \ldots$ Moreover,*

$$\|x_{n+1} - x_n\| \leq t_{n+1} - t_n \quad \text{and} \quad \|x^* - x_n\| \leq t^* - t_n, \quad n \geq 0,$$

where $\{t_n\}$ is defined in (4.1).

Proof. From (J1)-(J2) it is clear that $\|x_1 - x_0\| \leq t_1 - t_0 < t^* - t_0$ and $x_1 \in B(x_0, t^* - t_0) \subset \Omega$. If we now suppose that $x_j \in B(x_0, t^* - t_0)$, for $j = 1, 2, \ldots, n - 1$, from the last lemma, it follows that the operator Γ_{n-1} exists and $\|\Gamma_{n-1}\| \leq -\frac{1}{f'(t_{n-1})}$, so that x_n is well defined. Besides, since $\|x_{j+1} - x_j\| \leq t_{j+1} - t_j$ for $j = 1, 2, \ldots, n - 1$, we have that $\|x_n - x_0\| \leq t_n - t_0 < t^* - t_0$ and $x_n \in B(x_0, t^* - t_0)$. As a consequence, the sequence $\{x_n\}$ is well defined and $x_n \in B(x_0, t^* - t_0) \subset \Omega$ for all $n = 0, 1, 2, \ldots$

Moreover, since (4.1) is a majorizing sequence of Newton's sequence $\{x_n\}$ and convergent, then sequence $\{x_n\}$ is also convergent. As a consequence, $\lim_n x_n = x^* \in \overline{B(x_0, t^* - t_0)}$, where $t^* = \lim_n t_n$. From item (iii_n) of the last lemma, we have that $\|F(x_n)\| \leq f(t_n)$, for all $n = 0, 1, 2, \ldots$ Then, by the continuity of the operator F and letting $n \to +\infty$, we obtain that $F(x^*) = 0$.

Furthermore, for $j > 0$, we see

$$\|x_{n+j} - x_n\| \leq \sum_{i=1}^{j} \|x_{n+i} - x_{n+i-1}\| \leq \sum_{i=1}^{j} \|t_{n+i} - t_{n+i-1}\| = t_{n+j} - t_n,$$

and, by letting $j \to +\infty$, it follows that $\|x^* - x_n\| \leq t^* - t_n$, for all $n = 0, 1, 2, \ldots$ ∎

3.1.2 Uniqueness of solution and order of convergence

Next, we give a result on the uniqueness of solution. We emphasize that this result also generalizes that of Kantorovich under conditions (A1)-(A2)-(A3). Previously, we give the following necessary lemma.

Lemma 3.5. *Under the hypotheses of Theorem 3.4, if $x \in \overline{B(x_0, t^{**} - t_0)} \cap \Omega$, then $\|F''(x)\| \leq f''(t)$ when $\|x - x_0\| \leq t - t_0$.*

Proof. From the Taylor series, it follows

$$F''(x) = \sum_{j=2}^{k-1} \frac{1}{(j-2)!} F^{(j)}(x_0)(x-x_0)^{j-2} + \frac{1}{(k-3)!} \int_{x_0}^{x} F^{(k)}(z)(x-z)^{k-3} dz$$

$$= \sum_{j=2}^{k-1} \frac{1}{(j-2)!} F^{(j)}(x_0)(x-x_0)^{j-2}$$

$$+ \frac{1}{(k-3)!} \int_{0}^{1} F^{(k)}\left(x_0 + \tau(x-x_0)\right)(x-x_0)^{k-2}(1-\tau)^{k-3} d\tau.$$

Taking norms and the last lemma, $\|F^{(k)}(x)\| \leq \ell$, for $x \in \Omega$, and $f^{(k)}(t) = \ell$, we have

$$\|F''(x)\| \leq \sum_{j=2}^{k-1} \frac{1}{(j-2)!} \|F^{(j)}(x_0)\| \|x-x_0\|^{j-2}$$

$$+ \frac{1}{(k-3)!} \int_{x_0}^{x} \left\|F^{(k)}\left(x_0 + \tau(x-x_0)\right)\right\| (1-\tau)^{k-3} d\tau \|x-x_0\|^{k-2}$$

$$\leq \sum_{j=2}^{k-1} \frac{1}{(j-2)!} f^{(j)}(t_0)(t-t_0)^{j-2}$$

$$+ \frac{1}{(k-3)!} \int_{0}^{1} f^{(k)}\left(t_0 + \tau(t-t_0)\right)(1-\tau)^{k-3} d\tau \, (t-t_0)^{k-2}$$

$$= f''(t)$$

for $\|x-x_0\| \leq t - t_0$. ∎

Now, we are ready to give the following result on the uniqueness of solution.

Theorem 3.6. *Under the hypotheses of Theorem 3.4, the solution x^* is unique in $B(x_0, t^{**} - t_0) \cap \Omega$ if $t^* < t^{**}$ or in $\overline{B(x_0, t^* - t_0)}$ if $t^* = t^{**}$.*

Proof. First, we suppose that $t^* < t^{**}$ and y^* is another solution of $F(x) = 0$ in $B(x_0, t^{**} - t_0) \cap \Omega$. Then,

$$\|y^* - x_0\| \leq \rho(t^{**} - t_0) \quad \text{with} \quad \rho \in (0,1).$$

We now suppose that $\|y^* - x_j\| \leq \rho^{2^j}(t^{**} - t_j)$ for $j = 0, 1, \ldots, n$. In addition,

$$\|y^* - x_{n+1}\| = \|\Gamma_n\left(F(y^*) - F(x_n) - F'(x_n)(y^* - x_n)\right)\|$$

$$\leq \|\Gamma_n\| \int_{0}^{1} \|F''\left(x_n + \tau(y^* - x_n)\right)\| (1-\tau)\|y^* - x_n\|^2 d\tau.$$

As

$$\|x_n + \tau(y^* - x_n) - x_0\| \leq \|x_n - x_0\| + \tau\|y^* - x_n\| \leq t_n + \tau(t^{**} - t_n) - t_0,$$

it follows that

$$\int_{0}^{1} \|F''\left(x_n + \tau(y^* - x_n)\right)\| (1-\tau) d\tau \leq \int_{0}^{1} f''\left(t_n + \tau(t^{**} - t_n)\right)(1-\tau) d\tau \equiv \mu$$

from the previous lemma. As a consequence,

$$\|y^* - x_{n+1}\| \leq -\frac{\mu}{f'(t_n)} \|y^* - x_n\|^2.$$

On the other hand, we also have

$$t^{**} - t_{n+1} = -\frac{1}{f'(t_n)} \left(f(t^{**}) - f(t_n) - f'(t_n)(t^{**} - t_n) \right)$$

$$= -\frac{1}{f'(t_n)} \int_0^1 f'' \left(t_n + \tau(t^{**} - t_n) \right) (1 - \tau)(t^{**} - t_n)^2 \, d\tau$$

$$= -\frac{\mu}{f'(t_n)} (t^{**} - t_n)^2.$$

Therefore,

$$\|y^* - x_{n+1}\| \leq \frac{t^{**} - t_{n+1}}{(t^{**} - t_n)^2} \|y^* - x_n\|^2 \leq \frac{t^{**} - t_{n+1}}{(t^{**} - t_n)^2} \left(\rho^{2^n}(t^{**} - t_n) \right)^2 = \rho^{2^{n+1}}(t^{**} - t_{n+1}).$$

so that $y^* = x^*$, since $\lim_n x_n = x^*$.

Second, the case $t^* = t^{**}$ follows similarly to the previous one by proving now by induction that $\|y^* - x_n\| \leq t^{**} - t_n = t^* - t_n$ and taking into account that $\lim_n t_n = t^*$. ■

Finally, we see the quadratic convergence of Newton's method under conditions (J1)-(J2)-(J3). We obtain the following theorem from Ostrsowski's technique [65]. The proof of the theorem is completely analogous to that of Theorem 2.18.

Theorem 3.7. *Under the hypotheses of Theorem 3.4 and assuming that* $f(t) = (t^* - t)(t^{**} - t)g(t)$, *where* $g(t)$ *is a function such that* $g(t^*) \neq 0$ *and* $g(t^{**}) \neq 0$, *we have the following:*

(a) *If* $t^* < t^{**}$ *and there exist* $m_1 > 0$ *and* $M_1 > 0$ *such that* $m_1 \leq \min\{Q_1(t); t \in [t_0, t^*]\}$ *and* $M_1 \geq \max\{Q_1(t); t \in [t_0, t^*]\}$, *where* $Q_1(t) = \frac{(t^{**}-t)g'(t)-g(t)}{(t^*-t)g'(t)-g(t)}$, *then*

$$\frac{(t^{**} - t^*)\theta^{2^n}}{m_1 - \theta^{2^n}} \leq t^* - t_n \leq \frac{(t^{**} - t^*)\Delta^{2^n}}{M_1 - \Delta^{2^n}}, \quad n \geq 0,$$

where $\theta = \frac{t^*}{t^{**}}m_1$, $\Delta = \frac{t^*}{t^{**}}M_1$ *and provided that* $\theta < 1$ *and* $\Delta < 1$.

(b) *If* $t^* = t^{**}$ *and there exist* $m_2 > 0$ *and* $M_2 > 0$ *such that* $m_2 \leq \min\{Q_2(t); t \in [t_0, t^*]\}$ *and* $M_2 \geq \max\{Q_2(t); t \in [t_0, t^*]\}$, *where* $Q_2(t) = \frac{(t^*-t)g'(t)-g(t)}{(t^*-t)g'(t)-2g(t)}$, *then*

$$m_2^n t^* \leq t^* - t_n \leq M_2^n t^*, \quad n \geq 0,$$

provided that $m_2 < 1$ *and* $M_2 < 1$.

From the last theorem, it follows that the convergence of Newton's method, under conditions (J1)-(J2)-(J3) , is quadratic if $t^* < t^{**}$ and linear if $t^* = t^{**}$.

3.1.3 Applications

Now, we illustrate the study presented until now with two nonlinear conservative problems of form (1.44)-(1.45). For both problems, the study above mentioned improves those of Kantorovich under conditions (A1)-(A2)-(A3). In the first example, we improve the results obtained from the Newton-Kantorovich theorem, since the domains of existence and uniqueness of solution and the a priori error bounds are improved respectively from Theorems 3.4

and 3.6. Moreover, we also see that we can guarantee the semilocal convergence of Newton's method to a solution of the problem from Theorem 3.4, but not from the Newton-Kantorovich theorem. In the second example, we consider a problem where the Newton-Kantorovich theorem cannot be applied to guarantee the semilocal convergence of Newton's method to the solutions of the problem, but we can do it from Theorem 3.4.

In practice, we usually choose $t_0 = 0$, since function (3.3) has the same property of translation as Kantorovich's polynomial (see Remark 1.18) and majorizes the operator F, in the sense of Definition 1.26, independent of the value of t_0.

Example 3.8. Firstly, we consider the polynomial law given by (1.53) for the heat generation in the nonlinear conservative system given in (1.44)-(1.45); i.e.:

$$\frac{d^2x(s)}{ds^2} + 3 + x(s) + 2x(s)^2 + x(s)^3 = 0, \qquad x(0) = x(1) = 0. \tag{3.4}$$

In Section 1.2.2, we have seen that, after following a process of discretization, a discrete numerical solution $\mathbf{x}^*$ of the above problem is given in Table 1.2, along with the domains of existence and uniqueness of solution. In particular, the solution $\mathbf{x}^*$ given in Table 1.2 exists and is unique, respectively, in

$$\{\mathbf{v} \in \mathbb{R}^8 : \|\mathbf{v}\| \leq 0.7047\ldots\} \quad \text{and} \quad \{\mathbf{v} \in \mathbb{R}^8 : \|\mathbf{v}\| < 1.0015\ldots\} \cap \Lambda = \Lambda,$$

Remember that the solution $\mathbf{x}^*$ given in Table 1.2 is the only solution of the problem to which we can guarantee the convergence of Newton's method by the Newton-Kantorovich theorem.

In addition, if we choose $t_0 = 0$, for Theorems 3.4 and 3.6, we obtain $\|\mathbb{F}''(\mathbf{x_0})\| \leq \frac{4}{81} = b_2$, $\|\mathbb{F}'''(\mathbf{x})\| \leq \frac{2}{27} = \ell$, $f(t) = \frac{t^3}{81} + \frac{2}{81}t^2 - (0.0895\ldots)t + (0.0370\ldots)$, $\alpha = 1.0250\ldots$ and $f(\alpha) = -0.0154\ldots \leq 0$, so that the conditions of Theorems 3.4 and 3.6 hold. Besides, the positive real roots of $f(t) = 0$ are $t^* = 0.4997\ldots$ and $t^{**} = 1.5005\ldots$, and consequently the domains of existence and uniqueness of solution are respectively

$$\{\mathbf{v} \in \mathbb{R}^8 : \|\mathbf{v}\| \leq 0.4997\ldots\} \quad \text{and} \quad \{\mathbf{v} \in \mathbb{R}^8 : \|\mathbf{v}\| < 1.5005\ldots\},$$

so that the domains obtained from Theorems 3.4 and 3.6 improve those given by the Newton-Kantorovich theorem, since the domain of existence of solution is smaller and the domain of uniqueness of solution is bigger.

Also, we see that the corresponding majorizing sequence given by (4.1) provides better error estimates than those obtained from the majorizing sequence $\{s_n\}$ obtained from Kantorovich's polynomial $p(s) = \frac{17}{324}s^2 - (0.0895\ldots)s + (0.0370\ldots)$. The error estimates and the absolute error are shown in Table 3.1. Observe the remarkable improvement obtained from the majorizing real sequence $\{t_n\}$ constructed from the function f given above.

On the other hand, we have seen in Section 1.2.2 that problem (3.4) may have a solution $x^{**}(s)$ such that $\|x^{**}(s)\| \geq \rho_2 = 1.4291\ldots$ In this case, $\Omega = C^2([0,1])$ and $\Lambda = \mathbb{R}^8$. If we now choose, for example, the starting vector $\mathbf{x_0} = (2, 2, \ldots, 2)^T$ for Newton's method, we observe that $\|\mathbf{x_0}\| = 2 > \rho_2 = 1.4291\ldots$, so that we cannot apply Theorem 3.4 either, since $f(\alpha) = 0.1157\ldots > 0$ with $\alpha = 0.4952\ldots$ and

$$f(t) = \frac{t^3}{81} + \frac{8}{81}t^2 - (0.1069\ldots)t + (0.1430\ldots),$$

n	$\|\mathbf{x}^* - \mathbf{x_n}\|$	$|t^* - t_n|$	$|s^* - s_n|$
0	$4.6853\ldots \times 10^{-1}$	$4.9978\ldots \times 10^{-1}$	$7.0479\ldots \times 10^{-1}$
1	$5.4855\ldots \times 10^{-2}$	$8.6102\ldots \times 10^{-2}$	$2.9110\ldots \times 10^{-1}$
2	$1.1968\ldots \times 10^{-3}$	$4.8519\ldots \times 10^{-3}$	$9.6411\ldots \times 10^{-2}$
3	$6.0715\ldots \times 10^{-7}$	$1.8100\ldots \times 10^{-5}$	$1.8983\ldots \times 10^{-2}$
4	$1.5627\ldots \times 10^{-13}$	$2.5471\ldots \times 10^{-10}$	$1.0732\ldots \times 10^{-3}$

Table 3.1: Absolute errors and error estimates

$\beta = 9.3529\ldots$, $\eta = 1.3375\ldots$ and $b_2 = \frac{16}{81}$. However, it seems clear that the conditions of Theorem 3.4 can be satisfied if the starting point is improved. So, taking $\mathbf{x_0} = (2, 2, \ldots, 2)^T$, after two iterations of Newton's method, we obtain the vector

$$\mathbf{x_2} = \begin{pmatrix} 0.70391713\ldots \\ 1.34565183\ldots \\ 1.85928293\ldots \\ 2.15200198\ldots \\ 2.15200198\ldots \\ 1.85928293\ldots \\ 1.34565183\ldots \\ 0.70391713\ldots \end{pmatrix},$$

which is now used as new starting point for Newton's method, $\tilde{\mathbf{x}}_0 = \mathbf{x_2}$. For this new starting point, we obtain

$$f(t) = \frac{t^3}{81} + (0.1043\ldots)t^2 - (0.0874\ldots)t + 0.0039\ldots,$$

$\alpha = 0.3917\ldots$ and $f(\alpha) = -0.0135\ldots \leq 0$, so that the hypotheses of Theorem 3.4 are then satisfied for $\tilde{\mathbf{x}}_0$. As a consequence, $t^* = 0.0481\ldots$, $t^{**} = 0.7235\ldots$ and the domains of existence and uniqueness of solution are respectively

$$\{\mathbf{v} \in \mathbb{R}^8 : \|\mathbf{v} - \tilde{\mathbf{x}}_0\| \leq 0.0481\ldots\} \quad \text{and} \quad \{\mathbf{v} \in \mathbb{R}^8 : \|\mathbf{v} - \tilde{\mathbf{x}}_0\| < 0.7235\ldots\}.$$

Observe that $\tilde{\mathbf{x}}_0$ satisfies $\|\tilde{\mathbf{x}}_0\| = 2.1520\ldots > \rho_2 = 1.4291\ldots$ After four more iterations of Newton's method, we obtain the approximated solution $\mathbf{x}^{**} = (x_1^{**}, x_2^{**}, \ldots, x_8^{**})^T$ given in Table 3.2, which is a solution that is beyond the scope of the Newton-Kantorovich theorem. Observe that $\|\mathbf{x}^{**}\| = 2.1050\ldots > \rho_2 = 1.4291\ldots$

In Table 3.3 we show the errors $\|\mathbf{x}^{**} - \tilde{\mathbf{x}}_n\|$, the error estimates and the sequence $\{\|\mathbb{F}(\tilde{\mathbf{x}}_n)\|\}$. From the last, we notice that the vector shown in Table 3.2 is a good approximation of the solution of the nonlinear algebraic system given by (1.52)-(1.53) with $m = 8$.

Finally, by interpolating the values of Table 3.2 and taking into account the boundary conditions, we obtain the solution denoted by $\tilde{\tilde{x}}$ and drawn in Figure 3.1. Note that the solution denoted by $\tilde{x}$ is that obtained for the numerical solution $\mathbf{x}^*$ of problem (3.4) given in Table 1.2.

i	x_i^{**}	i	x_i^{**}
1	0.68796551...	5	2.10501296...
2	1.31469436...	6	1.81742466...
3	1.81742466...	7	1.31469436...
4	2.10501296...	8	0.68796551...

Table 3.2: Numerical solution $\mathbf{x}^{**}$ of (1.52) with $\phi(u)$ defined in (1.53)

n	$\|\mathbf{x}^{**} - \tilde{\mathbf{x}}_n\|$	$|t^* - t_n|$	$\|\mathbb{F}(\tilde{\mathbf{x}}_n)\|$
0	$4.6989\ldots \times 10^{-2}$	$4.8119\ldots \times 10^{-2}$	$8.2734\ldots \times 10^{-3}$
1	$1.6485\ldots \times 10^{-3}$	$2.7787\ldots \times 10^{-3}$	$2.1346\ldots \times 10^{-4}$
2	$2.0311\ldots \times 10^{-6}$	$1.0510\ldots \times 10^{-5}$	$2.7841\ldots \times 10^{-7}$
3	$3.1208\ldots \times 10^{-12}$	$1.5186\ldots \times 10^{-10}$	$4.2350\ldots \times 10^{-13}$

Table 3.3: Absolute errors, error estimates and $\{\|\mathbb{F}(\tilde{\mathbf{x}}_n)\|\}$

Example 3.9. Secondly, we choose the following polynomial law

$$\phi(x(s)) = x(s) + x(s)^3 \tag{3.5}$$

for the heat generation in the nonlinear conservative system given in (1.44)-(1.45); i.e.:

$$\frac{d^2 x(s)}{ds^2} + x(s) + x(s)^3 = 0, \qquad x(0) = x(1) = 0. \tag{3.6}$$

We are interested in other than the trivial solution. Note that a solution $x^*(s)$ of problem (3.6), as a consequence of a previous location, must satisfy in $\mathcal{C}^2([0,1])$ that $\|x^*(s)\| \geq \sqrt{7}$. In the following, we guarantee the convergence of Newton's method to other solution different from the trivial one from Theorem 3.4 by considering a suitable k, that in this case is $k = 3$.

We now choose $m = 8$, $\Lambda = \mathbb{R}^8$ and the starting point $\mathbf{x_0} = (3, 3, \ldots, 3)^T$, which is such that $\|\mathbf{x_0}\| = 3 > \sqrt{7}$. From this starting point, we cannot define the corresponding scalar function f for Theorem 3.4. But, if we apply Newton's method with $\mathbf{x_0}$ to solve (1.52) with $\phi(x)$ defined in (3.5), after two iterations, we obtain the vector

$$\mathbf{x_2} = \begin{pmatrix} 1.04992437\ldots \\ 2.07377016\ldots \\ 2.96195643\ldots \\ 3.50405170\ldots \\ 3.50405170\ldots \\ 2.96195643\ldots \\ 2.07377016\ldots \\ 1.04992437\ldots \end{pmatrix},$$

that can be considered as new starting vector for Newton's method, $\tilde{\mathbf{x}}_0 = \mathbf{x_2}$, so that the convergence of the method is now guaranteed from Theorem 3.4. For this new starting point,

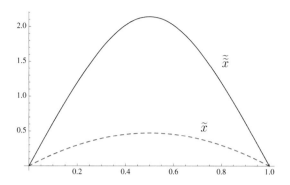

Figure 3.1: Approximated solutions of (1.44)-(1.45) with $\phi(u)$ defined in (1.53)

we obtain the scalar function

$$f(t) = \frac{t^3}{81} + (0.1544\ldots)\, t^2 - (0.1621\ldots)\, t + (0.0127\ldots),$$

with $t_0 = 0$, $\alpha = 0.4954\ldots$ and $f(\alpha) = -0.0281\ldots \leq 0$, so that the hypotheses of Theorem 3.4 are then satisfied. In addition, after four more iterations of Newton's method, we obtain the approximated solution $\mathbf{x}^* = (x_1^*, x_2^*, \ldots, x_8^*)^T$ given in Table 3.4, which is such that $\|\mathbf{x}^*\| = 3.4220\ldots > \sqrt{7}$. Moreover, $t^* = 0.0859\ldots$ and $t^{**} = 0.8933\ldots$, so that the domains of existence and uniqueness of solution are respectively

$$\{\mathbf{v} \in \mathbb{R}^8 : \|\mathbf{v} - \tilde{\mathbf{x}}_0\| \leq 0.0859\ldots\} \quad \text{and} \quad \{\mathbf{v} \in \mathbb{R}^8 : \|\mathbf{v} - \tilde{\mathbf{x}}_0\| < 0.8933\ldots\}$$

i	x_i^*	i	x_i^*
1	1.02086468...	5	3.42206801...
2	2.01599142...	6	2.88507585...
3	2.88507585...	7	2.01599142...
4	3.42206801...	8	1.02086468...

Table 3.4: Numerical solution $\mathbf{x}^*$ of (1.52) with $\phi(u)$ defined in (3.5)

In Table 3.5 we show the errors $\|\mathbf{x}^* - \tilde{\mathbf{x}}_n\|$, the error estimates and the sequence $\{\|\mathbb{F}(\tilde{\mathbf{x}}_n)\|\}$. From the last, we notice that the vector shown in Table 3.4 is a good approximation of the solution of the nonlinear algebraic system given by (1.52)-(3.5) with $m = 8$.

3.2 Operators with ω-bounded k-th-derivative

Return to the conservative problem introduced in Section 3.1. If the law $\phi(x(s))$, for the heat generation, is not polynomial, but other expressions are, such as, for example

$$\phi(x(s)) = 1 + x(s) + x(s)^2 + x(s)^3 + e^{x(s)}, \tag{3.7}$$

n	$\|\mathbf{x}^* - \widetilde{\mathbf{x}}_\mathbf{n}\|$	$|t^* - t_n|$	$\|\mathbb{F}(\widetilde{\mathbf{x}}_\mathbf{n})\|$
0	$8.1983\ldots \times 10^{-2}$	$8.5927\ldots \times 10^{-2}$	$3.2326\ldots \times 10^{-2}$
1	$3.1380\ldots \times 10^{-3}$	$7.0813\ldots \times 10^{-3}$	$8.0074\ldots \times 10^{-4}$
2	$4.1302\ldots \times 10^{-6}$	$5.7400\ldots \times 10^{-5}$	$1.2455\ldots \times 10^{-6}$
3	$7.5911\ldots \times 10^{-12}$	$3.8372\ldots \times 10^{-9}$	$2.1621\ldots \times 10^{-12}$

Table 3.5: Absolute errors, error estimates and $\{\|\mathbb{F}(\widetilde{\mathbf{x}}_\mathbf{n})\|\}$

the argument used in Section 3.1 to bound $\|\mathfrak{F}''(x)\|$ is not valid, since condition (J2) is not satisfied for any value of k. However, if we consider $\phi^{(iv)}(x(s))$, then $\phi^{(iv)}(x(s)) = e^{x(s)}$ and we can think of a condition like that given for $\|\mathfrak{F}''(x)\|$ in (2.30), Section 2.2. In this case,

$$\|\mathfrak{F}^{(iv)}(x)\| \leq \omega(\|x\|), \quad x \in \Omega,$$

where $\omega : [0, +\infty) \longrightarrow \mathbb{R}$ is a continuous monotonous (nondecreasing or nonincreasing) function. This generalization to high-order derivatives has the advantage that allow not having to locate previously the solution in order to find a domain where $\|\mathfrak{F}^{(iv)}(x)\|$ is bounded.

The rest of Section 3.2 is divided in five subsections. In Section 3.2.1 we recall the concept of majorizing sequence and introduce the new convergence conditions. In Section 3.2.2, we prove the existence of a majorant function. In Section 3.2.3, we establish a general semilocal convergence result for Newton's method, under the new convergence conditions introduced previously, which includes information about the existence of solution, and indicate how the majorizing sequence is constructed. We include information about the uniqueness of solution and a result on the a priori error estimates that leads to the quadratic convergence of Newton's method in Section 3.2.4. Finally, in Section 3.2.5, we present an application where a conservative problem is involved. We clearly show the advantages of the semilocal convergence result established previously with respect to the Newton-Kantorovich theorem.

3.2.1 Existence of a majorizing sequence

As we have seen previously, condition (A2) limits the application of the Newton-Kantorovich theorem. The aim of this section is to generalize Kantorovich's conditions by modifying conditions (a)-(b) of Theorem 1.27. For this, we construct a scalar function $f \in \mathcal{C}^k([t_0, +\infty))$, with $t_0 \in \mathbb{R}$, $k \geq 3$, that satisfies:

(K1) There exists $\Gamma_0 = [F'(x_0)]^{-1} \in \mathcal{L}(Y, X)$, for some $x_0 \in \Omega$, with $\|\Gamma_0\| \leq -\frac{1}{f'(t_0)}$ and $\|\Gamma_0 F(x_0)\| \leq -\frac{f(t_0)}{f'(t_0)}$; moreover, $\|F^{(i)}(x_0)\| \leq f^{(i)}(t_0)$, with $i = 2, 3, \ldots, k-1$ and $k \geq 3$.

(K2) $\|F^{(k)}(x)\| \leq f^{(k)}(t)$, for $\|x - x_0\| \leq t - t_0$, for $x \in \Omega$ and $t \in [t_0, +\infty)$.

Next, we prove the semilocal convergence of Newton's method under conditions (K1)-(K2), by using the modification of the majorant principle of Kantorovich introduced in Section 1.1.4, as we have done under conditions (a)-(b) of Theorem 1.27 (case $k = 2$ in (K2)). So, we construct again a majorizing real sequence $\{t_n\}$, as that given in (4.1), for Newton's sequence $\{x_n\}$ in the Banach space X.

As we can see in the Newton-Kantorovich theorem, the majorizing real sequence $\{s_n\}$ that is constructed from Newton's method and Kantorovich's polynomial (1.23) converges to the smallest positive solution s^* of the equation $p(s) = 0$. This convergence is clear, since Kantorovich's polynomial (1.23) is a nonincreasing and convex function in $[0, s^*]$. To apply the modification of the majorant principle of Kantorovich to our particular situation, the equation $f(t) = 0$ must have at least one solution t^* greater than t_0, so that the corresponding majorizing real sequence $\{t_n\}$, under conditions (K1)-(K2), converges to t^* from t_0. Clearly, the first thing we need is to study the function f. Then, we give some properties of this function.

Theorem 3.10. *Let $f \in \mathcal{C}^k([t_0, +\infty))$ be with $t_0 \in \mathbb{R}$, $k \geq 3$, such that $f(t_0) > 0$, $f'(t_0) < 0$, $f^{(i)}(t_0) > 0$, for $i = 2, 3, \ldots, k-1$, and $f^{(k)}(t) > 0$.*

    (a) If there exists a solution α of $f'(t) = 0$ such that $\alpha > t_0$, then α is the unique minimum of $f(t)$ in $[t_0, +\infty)$ and $f(t)$ is nonincreasing in $[t_0, \alpha)$.

    (b) If $f(\alpha) \leq 0$, then $f(t) = 0$ has at least one solution in $[t_0, +\infty)$. Moreover, if t^* is the smallest solution of $f(t) = 0$ in $[t_0, +\infty)$, then $t_0 < t^* \leq \alpha$.

Proof. First, as $f^{(k)}(t) > 0$, for all $t \in [t_0, +\infty)$, then $f^{(k-1)}(t)$ is nondecreasing, for all $t \in [t_0, +\infty)$. In addition, since $f^{(k-1)}(t_0) > 0$, then $f^{(k-1)}(t) > 0$, for all $t \in [t_0, +\infty)$. As a consequence, $f^{(k-2)}(t)$ is nondecreasing, for all $t \in [t_0, +\infty)$. Moreover, taking into account that $f^{(k-2)}(t_0) > 0$, it follows then $f^{(k-2)}(t) > 0$, for all $t \in [t_0, +\infty)$. Now, we repeat this procedure until $f''(t) > 0$, for all $t \in [t_0, +\infty)$ and, consequently, f is convex in $[t_0, +\infty)$. Besides, as $f'(\alpha) = 0$ with $\alpha > t_0$, then α is the unique minimum of f in $[t_0, +\infty)$.

    Second, $f'(t)$ is nondecreasing in $t \in [t_0, +\infty)$, since $f''(t) > 0$ in $t \in [t_0, +\infty)$. Moreover, as $f'(t_0) < 0$ and $f'(\alpha) = 0$, then $t_0 < \alpha$ and $f'(t) < 0$ in $[t_0, \alpha)$. Therefore, f is nonincreasing in $[t_0, \alpha)$.

    Third, if $f(\alpha) < 0$, then f has at least one zero t^* in (t_0, α), since $f(t_0) > 0$. In addition, as f nonincreasing in $[t_0, \alpha)$, t^* is the unique solution of $f(t) = 0$ in $[t_0, \alpha)$.

    Finally, if $f(\alpha) = 0$, then α is a double zero of f and $t^* = \alpha$. ∎

As we are interested in the fact that (4.1) is a majorizing real sequence of Newton's sequence $\{x_n\}$, defined in the Banach space X, we establish the convergence of $\{t_n\}$ in the next result.

Theorem 3.11. *Let $\{t_n\}$ be the scalar sequence given in (4.1) with $f \in \mathcal{C}^k([t_0, +\infty))$ and $t_0 \in \mathbb{R}$, $k \geq 3$. Suppose that conditions (K1)-(K2) hold and there exists a solution $\alpha \in (t_0, +\infty)$ of $f'(t) = 0$ such that $f(\alpha) \leq 0$. Then, (4.1) is a nondecreasing real sequence that converges to t^*.*

    Observe that if there exists such function f, then it has the same properties as Kantorovich's polynomial (1.23), since $f(t) > 0$, $f'(t) < 0$ and $f''(t) > 0$ in (t_0, α), so that the convergence of the real sequence given in (4.1) is then guaranteed in a way analogous to that of sequence (1.22) and such as we have seen in Theorem 1.21.

    Next, from the following lemma, we see that (4.1) is a majorizing sequence of Newton's sequence $\{x_n\}$ defined in the Banach space X.

Lemma 3.12. *Suppose that there exists $f \in \mathcal{C}^k([t_0, +\infty))$, with $t_0 \in \mathbb{R}$, such that conditions (K1)-(K2) are satisfied. Suppose also that $x_n \in \Omega$, for all $n \geq 0$, and $f(\alpha) \leq 0$, where α is a solution of $f'(t) = 0$ in $(t_0, +\infty)$. Then, for all $n \in \mathbb{N}$, we have the following:*

(i_n) *There exists* $\Gamma_n = [F'(x_n)]^{-1}$ *and* $\|\Gamma_n\| \leq -\dfrac{1}{f'(t_n)}$,

(ii_n) $\|F^{(i)}(x_n)\| \leq f^{(i)}(t_n)$, *for* $i = 2, 3, \ldots, k-1$,

(iii_n) $\|F(x_n)\| \leq f(t_n)$,

(iv_n) $\|x_{n+1} - x_n\| \leq t_{n+1} - t_n$.

Proof. We prove (i_n)-(ii_n)-(iii_n)-(iv_n) by mathematical induction on n. As the cases $n = 1$ and the inductive step are proved in the same way, we only write the inductive step. We then suppose that (i_n)-(ii_n)-(iii_n)-(iv_n) are true for $n = 1, 2, \ldots, d-1$ and prove that (i_d)-(ii_d)-(iii_d)-(iv_d) are also true.

First, from Taylor's series we have

$$I - \Gamma_{d-1}F'(x_d) = I - \Gamma_{d-1}\left(\sum_{j=1}^{k-1} \frac{1}{(j-1)!}F^{(j)}(x_{d-1})(x_d - x_{d-1})^{j-1}\right.$$
$$\left. + \frac{1}{(k-2)!}\int_{x_{d-1}}^{x_d} F^{(k)}(x)(x_d - x)^{k-2}\, dx\right).$$

If we now denote $x = x_{d-1} + \tau(x_d - x_{d-1})$ and $t = t_{d-1} + \tau(t_d - t_{d-1})$ with $\tau \in [0, 1]$, then

$$\|x - x_0\| \leq \tau\|x_d - x_{d-1}\| + \|x_{d-1} - x_{d-2}\| + \cdots + \|x_1 - x_0\|$$
$$\leq \tau(t_d - t_{d-1}) + t_{d-1} - t_{d-2} + \cdots + t_1 - t_0$$
$$= t - t_0.$$

Thus,

$$
\begin{aligned}
\|I - \Gamma_{d-1} F'(x_d)\| &\leq \|\Gamma_{d-1}\| \left(\sum_{j=2}^{k-1} \frac{1}{(j-1)!} \|F^{(j)}(x_{d-1})\| \|x_d - x_{d-1}\|^{j-1} \right. \\
&\quad + \frac{1}{(k-2)!} \int_0^1 \left\| F^{(k)}(x_{d-1} + \tau(x_d - x_{d-1})) \right\| \\
&\quad \left. \times (1-\tau)^{k-2} d\tau \, \|x_d - x_{d-1}\|^{k-1} \right) \\
&\leq -\frac{1}{f'(t_{d-1})} \left(\sum_{j=2}^{k-1} \frac{1}{(j-1)!} f^{(j)}(t_{d-1})(t_d - t_{d-1})^{j-1} \right. \\
&\quad + \frac{1}{(k-2)!} \int_0^1 f^{(k)}(t_{d-1} + \tau(t_d - t_{d-1})) \\
&\quad \left. \times (1-\tau)^{k-2} d\tau \, (t_d - t_{d-1})^{k-1} \right) \\
&= -\frac{1}{f'(t_{d-1})} \left(\sum_{j=2}^{k-1} \frac{1}{(j-1)!} f^{(j)}(t_{d-1})(t_d - t_{d-1})^{j-1} \right. \\
&\quad \left. + \frac{1}{(k-2)!} \int_{t_{d-1}}^{t_d} f^{(k)}(t)(t_d - t)^{k-2} \, dt \right) \\
&= -\frac{1}{f'(t_{d-1})} \left(f'(t_d) - f'(t_{d-1}) \right) \\
&= 1 - \frac{f'(t_d)}{f'(t_{d-1})} \\
&< 1,
\end{aligned}
$$

since $\frac{f'(t_d)}{f'(t_{d-1})} \in (0,1)$ to be $t_{d-1} < t_d \leq t^*$ and $f''(t) > 0$ in $[t_0, +\infty)$. As a consequence, by the Banach lemma on invertible operators, there exists Γ_d and $\|\Gamma_d\| \leq -\frac{1}{f'(t_d)}$.

Second, for any $i \in \{2, 3, \ldots, k-1\}$ and Taylor's series, we have

$$
\begin{aligned}
F^{(i)}(x_d) &= F^{(i)}(x_{d-1}) + F^{(i+1)}(x_{d-1})(x_d - x_{d-1}) + \frac{1}{2!} F^{(i+2)}(x_{d-1})(x_d - x_{d-1})^2 \\
&\quad + \cdots + \frac{1}{(k-1-i)!} F^{(k-1)}(x_{d-1})(x_d - x_{d-1})^{k-1-i} \\
&\quad + \frac{1}{(k-1-i)!} \int_{x_{d-1}}^{x_d} F^{(k)}(x)(x_d - x)^{k-1-i} \, dx.
\end{aligned}
$$

Moreover, as $\|x - x_0\| \le t - t_0$ and $\|x_d - x_{d-1}\| \le t_d - t_{d-1}$, it follows

$$
\begin{aligned}
\|F^{(i)}(x_d)\| &= \sum_{j=1}^{k-i} \frac{1}{(j-1)!} \|F^{(j+i-1)}(x_{d-1})\| \|x_d - x_{d-1}\|^{j-1} \\
&\quad + \frac{1}{(k-1-i)!} \int_0^1 \left\| F^{(k)}(x_{d-1} + \tau(x_d - x_{d-1})) \right\| \\
&\quad \times (1-\tau)^{k-2} d\tau \, \|x_d - x_{d-1}\|^{k-i} \\
&\le \sum_{j=1}^{k-i} \frac{1}{(j-1)!} f^{(j+i-1)}(t_{d-1})(t_d - t_{d-1})^{j-1} \\
&\quad + \frac{1}{(k-1-i)!} \int_0^1 f^{(k)}(t_{d-1} + \tau(t_d - t_{d-1})) \\
&\quad \times (1-\tau)^{k-2} d\tau \, (t_d - t_{d-1})^{k-i} \\
&= \sum_{j=1}^{k-i} \frac{1}{(j-1)!} f^{(j+i-1)}(t_{d-1})(t_d - t_{d-1})^{j-1} \\
&\quad + \frac{1}{(k-1-i)!} \int_{t_{d-1}}^{t_d} f^{(k)}(t)(t_d - t)^{k-i-1} dt \\
&= f^{(i)}(t_d).
\end{aligned}
$$

Third, from Taylor's series, we have

$$
\begin{aligned}
F(x_d) &= \sum_{j=0}^{k-1} \frac{1}{j} F^{(j)}(x_{d-1})(x_d - x_{d-1})^j \\
&\quad + \frac{1}{(k-1)!} \int_{x_{d-1}}^{x_d} F^{(k)}(z)(x_d - z)^{k-1} dz \\
&= \sum_{j=2}^{k-1} \frac{1}{j!} F^{(j)}(x_{d-1})(x_d - x_{d-1})^j \\
&\quad + \frac{1}{(k-1)!} \int_0^1 F^{(k)}(x_{d-1} + \tau(x_d - x_{d-1})) \\
&\quad \times (x_d - x_{d-1})^k (1-\tau)^{k-1} d\tau,
\end{aligned}
$$

since $F(x_{d-1}) + F'(x_{d-1})(x_d - x_{d-1}) = 0$. As a consequence,

$$
\begin{aligned}
\|F(x_d)\| &\leq \sum_{j=2}^{k-1} \frac{1}{j!} \|F^{(j)}(x_{d-1})\| \|x_d - x_{d-1}\|^j \\
&\quad + \frac{1}{(k-1)!} \int_0^1 \left\| F^{(k)}(x_{d-1} + \tau(x_d - x_{d-1})) \right\| \\
&\quad \times (1-\tau)^{k-1} d\tau \, \|x_d - x_{d-1}\|^k \\
&\leq \sum_{j=2}^{k-1} \frac{1}{j!} f^{(j)}(t_{d-1})(t_d - t_{d-1})^j \\
&\quad + \frac{1}{(k-1)!} \int_0^1 f^{(k)}(t_{d-1} + \tau(t_d - t_{d-1})) \\
&\quad \times (1-\tau)^{k-1}(t_d - t_{d-1})^k \, d\tau \\
&= \sum_{j=0}^{k-1} \frac{1}{j!} f^{(j)}(t_{d-1})(t_d - t_{d-1})^j \\
&\quad + \frac{1}{(k-1)!} \int_{t_{d-1}}^{t_d} f^{(k)}(z)(t_d - z)^{k-1} \, dz \\
&= f(t_d).
\end{aligned}
$$

Finally, $\|x_{d+1} - x_d\| \leq \|\Gamma_d\| \|F(x_d)\| \leq -\dfrac{f(t_d)}{f'(t_d)} = t_{d+1} - t_d$. ∎

Remark 3.13. From (K1)-(K2), items (i_0), (ii_0) and (iv_0) are obvious. Item (iii_0) does not need to prove (iv_0), since it follows from the condition $\|\Gamma_0 F(x_0)\| \leq \eta$.

Once we have seen that $\{t_n\}$ is a majorizing real sequence of the sequence $\{x_n\}$ defined in the Banach space X, we are ready to prove the semilocal convergence of $\{x_n\}$.

Theorem 3.14. (General semilocal convergence) *Let* $F : \Omega \subseteq X \longrightarrow Y$ *be a* k *(*$k \geq 3$*) times continuously Fréchet differentiable operator defined on a non-empty open convex domain* Ω *of a Banach space* X *with values in a Banach space* Y. *Suppose that there exist* $f \in \mathcal{C}^k([t_0, +\infty))$, *with* $t_0 \in \mathbb{R}$, *such that (K1)-(K2) are satisfied. If there exists* $\alpha \in (t_0, +\infty)$ *such that* $f'(\alpha) = 0$ *and* $f(\alpha) \leq 0$, *and* $B(x_0, t^* - t_0) \subset \Omega$. *Then, Newton's sequence* $\{x_n\}$ *converges to a solution* x^* *of* $F(x) = 0$ *starting at* x_0. *Moreover,* $x_n, x^* \in \overline{B(x_0, t^* - t_0)}$ *and*

$$\|x^* - x_n\| \leq t^* - t_n, \quad \text{for all} \quad n = 0, 1, 2, \ldots,$$

where $\{t_n\}$ *is defined in (4.1).*

Proof. From (K1)-(K2), it is clear that x_1 is well-defined and $\|x_1 - x_0\| = t_1 - t_0 < t^* - t_0$, so that $x_1 \in B(x_0, t^* - t_0) \subset \Omega$. Next, we suppose that x_j is well-defined and $x_j \in B(x_0, t^* - t_0)$ for $j = 1, 2, \ldots, n-1$.

Then, by the last lemma, the operator Γ_{n-1} exists and $\|\Gamma_{n-1}\| \leq -\dfrac{1}{f'(t_{n-1})}$. In addition, x_n is well-defined. Moreover, since $\|x_{j+1} - x_j\| \leq t_{j+1} - t_j$, for $j = 1, 2, \ldots, n-1$, we have

$$\|x_n - x_0\| \leq \sum_{i=0}^{n-1} \|x_{i+1} - x_i\| \leq \sum_{i=0}^{n-1}(t_{i+1} - t_i) = t_n - t_0 < t^* - t_0,$$

so that $x_n \in B(x_0, t^* - t_0)$. Therefore, Newton's sequence $\{x_n\}$ is well-defined and $x_n \in B(x_0, t^* - t_0)$ for all $n \geq 0$.

By the last lemma, we also have that $\|F(x_n)\| \leq f(t_n)$ and

$$\|x_{n+1} - x_n\| \leq \|\Gamma_n\| \|F(x_n)\| \leq -\frac{f(t_n)}{f'(t_n)} = t_{n+1} - t_n,$$

for all $n \geq 0$, so that $\{t_n\}$ is a majorizing real sequence of $\{x_n\}$ and convergent. Moreover, as $\lim_n t_n = t^*$, if $x^* = \lim_n x_n$, then $\|x^* - x_n\| \leq t^* - t_n$, for all $n = 0, 1, 2, \ldots$ Furthermore, as $\|F(x_n)\| \leq f(t_n)$, for all $n = 0, 1, 2, \ldots$, then by letting $n \to +\infty$, it follows $F(x^*) = 0$ by the continuity of F. $\blacksquare$

3.2.2   Existence of a majorant function

Until now, we have supposed that there exists $f \in C^k([t_0, +\infty))$, with $t_0 \in \mathbb{R}$, such that (K1)-(K2) are satisfied. Now, we see that this function exists and can then give a semilocal convergence result for Newton's method. For this, we suppose the following conditions:

(E1) There exists $\Gamma_0 = [F'(x_0)]^{-1} \in \mathcal{L}(Y, X)$, for some $x_0 \in \Omega$, with $\|\Gamma_0\| \leq \beta$ and $\|\Gamma_0 F(x_0)\| \leq \eta$; moreover, $\|F^{(i)}(x_0)\| \leq b_i$, with $i = 2, 3, \ldots, k - 1$ and $k \geq 3$.

(E2) There exists a continuous and nondecreasing function $\omega : [0, +\infty) \longrightarrow \mathbb{R}$ such that $\|F^{(k)}(x)\| \leq \omega(\|x\|)$ for $\|x - x_0\| \leq t - t_0$ and $\omega(0) = 0$.

Observe that we cannot obtain a real function by interpolation fitting under conditions (E1)-(E2), since (E2) does not allow determining the class of functions where (E1) can be applied.

Next, taking into account that $\|x\| \leq \|x_0\| + \|x - x_0\|$, from (E2), it follows

$$\|F^{(k)}(x)\| \leq \omega(\|x\|) \leq \omega(t - t_0 + \|x_0\|).$$

From now on, we denote $\varpi(t; \|x_0\|, t_0) = \omega(t - t_0 + \|x_0\|)$. So, condition (E2) is written as

$$\|F^{(k)}(x)\| \leq \varpi(t; \|x_0\|, t_0), \quad \text{for} \quad \|x - x_0\| \leq t - t_0,$$

where $\varpi : [t_0, +\infty) \longrightarrow \mathbb{R}$ is a continuous nondecreasing function such that $\varpi(t_0; \|x_0\|, t_0) \geq 0$.

Notice that the case in which the function ω is nonincreasing can be analyzed as that given in Section 2.2 for function (2.30).

Now, to find a real function f from conditions (E1)-(E2), it is enough to solve the following initial value problem:

$$\begin{cases} y^{(k)}(t) = \varpi(t; \|x_0\|, t_0), \\ y(t_0) = \dfrac{\eta}{\beta}, \quad y'(t_0) = -\dfrac{1}{\beta}, \\ y''(t_0) = b_2, \quad y'''(t_0) = b_3, \quad \ldots, \quad y^{(k-1)}(t_0) = b_{k-1}, \end{cases}$$

whose solution function is given in the next result.

Theorem 3.15. *For any nonnegative real numbers $\beta \neq 0$, η, $b_2, b_3, \ldots, b_{k-1}$, there exists only one solution $f(t)$ of the last initial value problem in $[t_0, +\infty)$; that is:*

$$f(t) = \int_{t_0}^{t} \int_{t_0}^{\theta_{k-1}} \cdots \int_{t_0}^{\theta_1} \varpi(\xi; \|x_0\|, t_0) \, d\xi \, d\theta_1 \cdots d\theta_{k-1}$$

$$+ \frac{b_{k-1}}{(k-1)!}(t-t_0)^{k-1} + \cdots + \frac{b_2}{2!}(t-t_0)^2 - \frac{t-t_0}{\beta} + \frac{\eta}{\beta}. \tag{3.8}$$

Remark 3.16. Observe that polynomial (3.3) obtained in Section 3.1 from interpolation fitting, can also be obtained from an initial value problem as above if ϖ is constant (see also [3, 27]).

To apply Theorem 4.4, the equation $f(t) = 0$ must have at least one solution greater than t_0, so that we have to guarantee the convergence of the scalar sequence $\{t_n\}$, from t_0, to this solution. We then study the function $f(t)$ defined in (3.8).

Theorem 3.17. *Let f be the function defined respectively in (3.8).*

(a) There exists only one positive solution $\alpha > t_0$ of the equation $f'(t) = 0$, which is the unique minimum of $f(t)$ in $[t_0, +\infty)$ and $f(t)$ is nonincreasing in $[t_0, \alpha)$.

(b) If $f(\alpha) \leq 0$, then the equation $f(t) = 0$ has at least one solution in $[t_0, +\infty)$. Moreover, if t^* is the smallest solution of $f(t) = 0$ in $[t_0, +\infty)$, we have $t_0 < t^* \leq \alpha$.

The proof of the last theorem is immediate from Theorem 3.10, since function (3.8) satisfies conditions (K1)-(K2).

3.2.3 Semilocal convergence

Taking into account the hypotheses of Theorem 3.17, function (3.8) satisfies the conditions of Theorem 4.4 and the semilocal convergence of Newton's method is then guaranteed in the Banach space X. In particular, we have the following theorem, whose proof follows immediately from Theorem 4.4. Previously, we consider:

(E3) $f(\alpha) \leq 0$, where f is the function given in (3.8) and α is the unique positive solution of $f'(t) = 0$ such that $\alpha > t_0$, and $B(x_0, t^* - t_0) \subset \Omega$, where t^* is the smallest positive solution of $f(t) = 0$ in $[t_0, +\infty)$.

Theorem 3.18. *Let $F : \Omega \subseteq X \longrightarrow Y$ be a k $(k \geq 3)$ times continuously Fréchet differentiable operator defined on a non-empty open convex domain Ω of a Banach space X with values in a Banach space Y and $f(t)$ the function defined in (3.8). Suppose that (E1)-(E2)-(E3) hold. Then, Newton's sequence $\{x_n\}$ converges to a solution x^* of $F(x) = 0$ starting at x_0. Moreover, $x_n, x^* \in \overline{B(x_0, t^* - t_0)}$ and*

$$\|x^* - x_n\| \leq t^* - t_n, \quad \text{for all} \quad n = 0, 1, 2, \ldots,$$

where $t_n = N_f(t_{n-1})$, with $n \in \mathbb{N}$.

3.2.4   Uniqueness of solution and order of convergence

After proving the semilocal convergence of Newton's method in X and locating the solution x^* of $F(x) = 0$, we prove the uniqueness of x^*. Before, we give the following technical lemma that is used later.

Lemma 3.19. *Under the hypotheses of Theorem 4.4, if the scalar function $f(t)$, given in (3.8), has two real zeros t^* and t^{**} such that $t_0 < t^* \leq t^{**}$ and $x \in \overline{B(x_0, t^{**} - t_0)} \cap \Omega$, then*

$$\|F''(x)\| \leq f''(t), \quad for \quad \|x - x_0\| \leq t - t_0.$$

Proof. From Taylor's series, it follows

$$F''(x) = \sum_{j=2}^{k-1} \frac{1}{(j-2)!} F^{(j)}(x_0)(x - x_0)^{j-2} + \frac{1}{(k-3)!} \int_{x_0}^x F^{(k)}(z)(x - z)^{k-3} dz$$

$$= \sum_{j=2}^{k-1} \frac{1}{(j-2)!} F^{(j)}(x_0)(x - x_0)^{j-2}$$

$$+ \frac{1}{(k-3)!} \int_0^1 F^{(k)}(x_0 + \tau(x - x_0))(1 - \tau)^{k-3}(x - x_0)^{k-2} d\tau.$$

If we take norms and take into account (K1)-(K2), for $\|x - x_0\| \leq t - t_0$, we have

$$\|F''(x)\| \leq \sum_{j=2}^{k-1} \frac{1}{(j-2)!} \|F^{(j)}(x_0)\| \|x - x_0\|^{j-2}$$

$$+ \frac{1}{(k-3)!} \int_0^1 \left\| F^{(k)}(x_0 + \tau(x - x_0)) \right\| (1 - \tau)^{k-3} d\tau \|x - x_0\|^{k-2}$$

$$\leq \sum_{j=2}^{k-1} \frac{1}{(j-2)!} f^{(j)}(t_0)(t - t_0)^{j-2}$$

$$+ \frac{1}{(k-3)!} \int_0^1 f^{(k)}(t_0 + \tau(t - t_0))(1 - \tau)^{k-3} d\tau (t - t_0)^{k-2}$$

$$= \sum_{j=2}^{k-1} \frac{1}{(j-2)!} f^{(j)}(t_0)(t - t_0)^{j-2} + \frac{1}{(k-3)!} \int_{t_0}^t f^{(k)}(s)(t - s)^{k-3} ds$$

$$= f''(t),$$

since $z = x_0 + \tau(x - x_0)$ and $s = t_0 + \tau(t - t_0)$ with $\tau \in [0, 1]$ and $\|z - x_0\| \leq \tau \|x - x_0\| \leq \tau - (t - t_0) = s - t_0$. $\blacksquare$

After that, we note that if $f(t)$ has two real zeros t^* and t^{**} such that $t_0 < t^* \leq t^{**}$, then the uniqueness of the solution follows from the next theorem, whose proof follows similarly to that given for Theorem 3.6.

Theorem 3.20. *Under the hypotheses of Theorem 4.4, if the scalar function $f(t)$ has two real zeros t^* and t^{**} such that $t_0 < t^* \leq t^{**}$, then the solution x^* is unique in $B(x_0, t^{**} - t_0) \cap \Omega$ if $t^* < t^{**}$ or in $\overline{B(x_0, t^* - t_0)}$ if $t^{**} = t^*$.*

Notice that the last theorem is a generalization of that obtained under Kantorovich's conditions.

Next, we see the quadratic convergence of Newton's method under conditions (K1)-(K2). Notice first that if $f(t)$ has two real zeros t^* and t^{**} such that $t_0 < t^* \leq t^{**}$, we can then write

$$f(t) = (t^* - t)(t^{**} - t)g(t)$$

with $g(t^*) \neq 0$ and $g(t^{**}) \neq 0$. Next, from Ostrsowski's technique [65], we give a theorem which provides some a priori error estimates for Newton's method. The proof of the theorem is completely analogous to that of Theorem 2.18.

Theorem 3.21. *Under the hypotheses of Theorem 4.4 and assuming that $f(t) = (t^* - t)(t^{**} - t)g(t)$, where $g(t)$ is a function such that $g(t^*) \neq 0$ and $g(t^{**}) \neq 0$, we have the following:*

(a) *If $t^* < t^{**}$ and there exist $m_1 > 0$ and $M_1 > 0$ such that $m_1 \leq \min\{Q_1(t); t \in [t_0, t^*]\}$ and $M_1 \geq \max\{Q_1(t); t \in [t_0, t^*]\}$, where $Q_1(t) = \frac{(t^{**} - t)g'(t) - g(t)}{(t^* - t)g'(t) - g(t)}$, then*

$$\frac{(t^{**} - t^*)\theta^{2^n}}{m_1 - \theta^{2^n}} \leq t^* - t_n \leq \frac{(t^{**} - t^*)\Delta^{2^n}}{M_1 - \Delta^{2^n}}, \quad n \geq 0,$$

where $\theta = \frac{t^*}{t^{**}} m_1$, $\Delta = \frac{t^*}{t^{**}} M_1$ and provided that $\theta < 1$ and $\Delta < 1$.

(b) *If $t^* = t^{**}$ and there exist $m_2 > 0$ and $M_2 > 0$ such that $m_2 \leq \min\{Q_2(t); t \in [t_0, t^*]\}$ and $M_2 \geq \max\{Q_2(t); t \in [t_0, t^*]\}$, where $Q_2(t) = \frac{(t^* - t)g'(t) - g(t)}{(t^* - t)g'(t) - 2g(t)}$, then*

$$m_2^n t^* \leq t^* - t_n \leq M_2^n t^*, \quad n \geq 0,$$

provided that $m_2 < 1$ and $M_2 < 1$.

From the last theorem, it follows that the convergence of Newton's method, under conditions (K1)-(K2), is quadratic if $t^* < t^{**}$ and linear if $t^* = t^{**}$.

3.2.5 Application

We illustrate the study developed above with a nonlinear conservative problem. This study improves that given by Kantorovich in the Newton-Kantorovich theorem. In particular, we improve the domains of existence and uniqueness of solution, along with the a priori error bounds. And, in addition, we see that the semilocal convergence of Newton's method to a solution of the problem can be guarantee from the above study, but not from the Newton-Kantorovich theorem.

Note that, in practice, we usually choose $t_0 = 0$, since function (3.8) has the same property of translation as Kantorovich's polynomial (see Remark 1.18) and majorizes the operator F, in the sense of Definition 1.26, independent of the value of t_0.

Example 3.22. We consider the particular case of (1.44)-(1.45), where the heat generation is now governed by the law defined in (3.7); i.e.:

$$\frac{d^2x(s)}{ds^2} + 1 + x(s) + x(s)^2 + x(s)^3 + e^{x(s)} = 0, \qquad x(0) = x(1) = 0. \qquad (3.9)$$

Following the process of discretization given in Section 1.2.2, we see that $v_\mathbf{x} = (1 + x_1 + x_1^2 + x_1^3 + e^{x_1}, \ldots, 1 + x_m + x_m^2 + x_m^3 + e^{x_m})^T$, where $\mathbf{x} = (x_1, x_2, \ldots, x_m)^T$. As a consequence, the first derivative of the function $\mathbb{F}$ defined in (1.52) is given by

$$\mathbb{F}'(\mathbf{x}) = A + h^2 D(d_\mathbf{x}),$$

where $d_\mathbf{x} = (1 + 2x_1 + 3x_1^2 + e^{x_1}, \ldots, 1 + 2x_m + 3x_m^2 + e^{x_m})^T$ and $D(d_\mathbf{x}) = \mathrm{diag}\{1 + 2x_1 + 3x_1^2 + e^{x_1}, \ldots, 1 + 2x_m + 3x_m^2 + e^{x_m}\}$. Moreover,

$$\mathbb{F}''(\mathbf{x})\,\mathbf{y}\,\mathbf{z} = (y_1, y_2, \ldots, y_m)\mathbb{F}''(\mathbf{x})(z_1, z_2, \ldots, z_m),$$

where $\mathbf{y} = (y_1, y_2, \ldots, y_m)^T$ and $\mathbf{z} = (z_1, z_2, \ldots, z_m)^T$, so that

$$\mathbb{F}''(\mathbf{x})\,\mathbf{y}\,\mathbf{z} = h^2((2 + 6x_1)y_1 z_1, (2 + 6x_2)y_2 z_2, \ldots, (2 + 6x_m)y_m z_m)^T.$$

Now, we study the application of the Newton-Kantorovich theorem to this problem. From (1.54) and

$$\|\mathbb{F}''(\mathbf{x})\mathbf{y}\,\mathbf{z}\| \le h^2 \left\| \begin{pmatrix} (2 + 6x_1 + e^{x_1})y_1 z_1 \\ (2 + 6x_2 + e^{x_2})y_2 z_2 \\ \vdots \\ (2 + 6x_m + e^{x_m})y_m z_m \end{pmatrix} \right\| \le h^2 \left(2 + 6\|\mathbf{x}\| + e^{\|x\|}\right) \|\mathbf{y}\|\|\mathbf{z}\|,$$

we observe that $\|\mathbb{F}''(\mathbf{x})\|$ is not bounded in general, since the scalar function deduced from the last expression and given by $\wp(t) = 2 + 6t + e^t$ is nondecreasing. Therefore, condition (A2) of Kantorovich is not satisfied, so that we cannot apply the Newton-Kantorovich theorem.

To solve the last difficulty and taking into account Remark 1.32, we look previously for a domain of location of solutions of the corresponding system of equations $\mathbb{F}(\mathbf{x}) = 0$ (see (1.52)) in order to apply condition (A2) of Kantorovich. For this, taking into account that the solution of problem (3.9) is of the form

$$x(s) = - \int_0^1 G(s, t)\phi(x(t))\, dt,$$

where $G(s, t)$ is the Green function in $[0, 1] \times [0, 1]$, we can locate the solution $x^*(s)$ in some domain. So, we have

$$\|x^*(s)\| - \frac{1}{8}\left(1 + \|x^*(2)\| + \|x^*(s)\|^2 + \|x^*(s)\|^3 + e^{\|x^*\|}\right) \le 0,$$

so that $\|x^*(s)\| \in [0, \rho_1] \cup [\rho_2, +\infty]$, where $\rho_1 = 0.3803\ldots$ and $\rho_2 = 1.4070\ldots$ are the two positive real roots of the scalar equation deduced from the last expression and given by $\chi(t) = t - \frac{1}{8}(1 + t + t^2 + t^3 + e^t) = 0$.

By the Newton-Kantorovich theorem, we could only guarantee the semilocal convergence of Newton's method to a solution $x^*(s)$ such that $\|x^*(s)\| \in [0, \rho_1]$. For this, we can consider the domain

$$\Omega = \left\{ x(s) \in \mathcal{C}^2([0, 1]) : \|x(s)\| < \frac{3}{4}, s \in [0, 1] \right\}$$

since $\rho_1 < \frac{3}{4} < \rho_2$.

In view of what the domain Ω is for (3.9), we then consider (1.52), $\mathbb{F}(\mathbf{x}) \equiv A\mathbf{x} + h^2 v_{\mathbf{x}} = 0$, with $\mathbb{F} : \Lambda \subset \mathbb{R}^m \longrightarrow \mathbb{R}^m$ and $\Lambda = \{\mathbf{x} \in \mathbb{R}^m : \|\mathbf{x}\| < \frac{3}{4}\}$, for the corresponding discrete problem.

If we choose $m = 8$ and the starting point $\mathbf{x_0} = (0, 0, \ldots, 0)^T$, we see that conditions of the Newton-Kantorovich theorem are satisfied, since

$$\|\mathbb{F}''(\mathbf{x})\| \le 0.1063 = M, \quad \beta = 12.6407\ldots, \quad \eta = 0.3121\ldots \quad \text{and} \quad M\beta\eta = 0.4197\ldots < \frac{1}{2}.$$

Moreover, as $p(s) = (0.0531\ldots)s^2 - (0.0791\ldots)s + (0.0246\ldots)$, then the two real zeros of p are $s^* = 0.4456\ldots$, $s^{**} = 1.0415\ldots$ and $B(\mathbf{x_0}, s^*) \subseteq \Lambda = B\left(0, \frac{3}{4}\right)$. Therefore, the domains of existence and uniqueness of solution are respectively

$$\{\mathbf{v} \in \mathbb{R}^8 : \|\mathbf{v}\| \le 0.4456\ldots\} \quad \text{and} \quad \{\mathbf{v} \in \mathbb{R}^8 : \|\mathbf{v}\| < 1.0415\ldots\} \cap \Lambda = \Lambda$$

and Newton's method converges to the solution $\mathbf{x}^* = (x_1^*, x_2^*, \ldots, x_8^*)^T$ shown in Table 3.6 after four iterations. Observe that $\|\mathbf{x}^*\| = 0.3365\ldots \le \frac{3}{4}$.

In Table 3.7 we show the errors $\|\mathbf{x}^* - \mathbf{x_n}\|$ and the sequence $\{\|\mathbb{F}(\mathbf{x_n})\|\}$. From the last, we notice that the vector shown in Table 3.6 is a good approximation of the solution of system (1.52)-(3.7) with $m = 8$.

i	x_i^*	i	x_i^*
1	0.12965132...	5	0.33657425...
2	0.23106719...	6	0.30091836...
3	0.30091836...	7	0.23106719...
4	0.33657425...	8	0.12965132...

Table 3.6: Numerical solution $\mathbf{x}^*$ of (1.52) with $\phi(u)$ defined in (3.7)

On the other hand, if we consider Theorem 3.18 with $t_0 = 0$, $\Omega = C^2([0, 1])$, $\Lambda = \mathbb{R}^8$ for the corresponding discrete problem and take into account that $\|\mathbb{F}''(\mathbf{x_0})\| \le \frac{1}{27} = b_2$ and $\varpi(t; \|x_0\|, t_0) = \varpi(t; 0, 0) = \frac{1}{81}(e^t + 6)$, we see that

$$f(t) = \frac{e^t}{81} + \frac{t^3}{81} + \frac{t^2}{81} - (0.0914\ldots)t + (0.0123\ldots),$$

$\alpha = 0.9706\ldots$ and $f(\alpha) = -0.0209\ldots < 0$. Therefore, the convergence of Newton's method to the solution $\mathbf{x}^*$ is also guaranteed from Theorem 3.18. Besides, the positive real roots of $f(t) = 0$ are $t^* = 0.3483\ldots$ and $t^{**} = 1.5007\ldots$, and consequently, the domains of existence and uniqueness of solution are respectively

$$\{\mathbf{v} \in \mathbb{R}^8 : \|\mathbf{v}\| \le 0.3483\ldots\} \quad \text{and} \quad \{\mathbf{v} \in \mathbb{R}^8 : \|\mathbf{v}\| < 1.5007\ldots\}.$$

We then see that the domains obtained from the Newton-Kantorovich theorem are improved by Theorems 3.18 and 4.5, since the domain of existence of solution is smaller and the domain of uniqueness of solution is bigger.

If we interpolate the values given in Table 3.7 and take into account the boundary conditions, we obtain the solution drawn in Figure 3.2 and denoted by $\tilde{x}$. Observe that $\|\mathbf{x}^*\| = 0.3365\ldots \le \frac{3}{4}$.

After that, we see that the corresponding majorizing real sequence $\{t_n\}$ provides better error estimates than those obtained from the majorizing real sequence $\{s_n\}$ constructed from Kantorovich's polynomial p. The error estimates and the absolute error are shown in Table 3.7. Observe the remarkable improvement obtained from the new majorizing sequence.

n	$\|x^* - x_n\|$	$|t^* - t_n|$	$|s^* - s_n|$	$\|F(x_n)\|$
0	$3.3657\ldots \times 10^{-1}$	$3.4831\ldots \times 10^{-1}$	$4.4566\ldots \times 10^{-1}$	$2.4691\ldots \times 10^{-2}$
1	$2.4455\ldots \times 10^{-1}$	$3.6197\ldots \times 10^{-2}$	$1.3354\ldots \times 10^{-1}$	$2.2471\ldots \times 10^{-3}$
2	$1.9682\ldots \times 10^{-4}$	$6.8108\ldots \times 10^{-4}$	$2.0665\ldots \times 10^{-2}$	$1.9248\ldots \times 10^{-5}$
3	$1.3106\ldots \times 10^{-8}$	$2.5867\ldots \times 10^{-7}$	$6.7015\ldots \times 10^{-4}$	$1.2956\ldots \times 10^{-8}$

Table 3.7: Absolute errors, error estimates and $\{\|F(x_n)\|\}$

Furthermore, we have seen previously that problem (3.9) may have a solution $x^{**}(s)$ such that $\|x^{**}(s)\| \geq \rho_2 = 1.4070\ldots$ We then guarantee the semilocal convergence of Newton's method from Theorem 3.18. In this case, $\Omega = \mathcal{C}^2([0, 1])$ and $\Lambda = \mathbb{R}^8$. If we now choose, for example, the starting vector $x_0 = (2, 2, \ldots, 2)^T$, we observe that $\|x_0\| = 2 > \rho_2 = 1.4070\ldots$ At first, we cannot apply Theorem 3.18 either, since $f(\alpha) = 0.1408\ldots > 0$ with $\alpha = 0.3897\ldots$ and

$$f(t) = (0.0912\ldots)e^t + \frac{t^3}{81} + \frac{7}{81}t^2 - (0.2076\ldots)t + (0.0731\ldots),$$

$\beta = 8.5868\ldots$, $\eta = 1.4117\ldots$ and $b_2 = 0.2640\ldots$ However, it seems clear that improving the initial approximation, the conditions of Theorem 3.18 hold. Indeed, applying Newton's method from $x_0 = (2, 2, \ldots, 2)^T$, after two iterations, we obtain the vector x_2, given by

$$x_2 = \begin{pmatrix} 0.67237222\ldots \\ 1.29091032\ldots \\ 1.78924172\ldots \\ 2.07401912\ldots \\ 2.07401912\ldots \\ 1.78924172\ldots \\ 1.29091032\ldots \\ 0.67237222\ldots \end{pmatrix},$$

that satisfies, as new starting point, the conditions of Theorem 3.18, since if we take $\tilde{x}_0 = x_2$, we obtain $\beta = 9.7742\ldots$, $\eta = 0.0618\ldots$, $b_2 = 0.2640\ldots$,

$$f(t) = (0.0912\ldots)e^t + (0.0123\ldots)t^3 + (0.0864\ldots)t^2 - (0.1935\ldots)t - (0.0848\ldots),$$

$\alpha = 0.3471\ldots$ and $f(\alpha) = -0.0120\ldots \leq 0$. Besides, $t^* = 0.0678\ldots$, $t^{**} = 0.6111\ldots$ and the domains of existence and uniqueness of solution are respectively

$$\{v \in \mathbb{R}^8 : \|u - \tilde{x}_0\| \leq 0.0678\ldots\} \quad \text{and} \quad \{v \in \mathbb{R}^8 : \|u - \tilde{x}_0\| < 0.6111\ldots\}.$$

Observe that the new starting point $\tilde{x}_0$ is such that $\|\tilde{x}_0\| = 2.0740\ldots > \rho_2 = 1.4070\ldots$ After four more iterations of Newton's method, we obtain the approximated solution $x^{**} =$

i	x_i^{**}	i	x_i^{**}
1	0.64954596...	5	2.00872701...
2	1.24649714...	6	1.72968114...
3	1.72968114...	7	1.24649714...
4	2.00872701...	8	0.64954596...

Table 3.8: Numerical solution $\mathbf{x}^{**}$ of (1.52) with $\phi(u)$ defined in (3.7)

$(x_1^{**}, x_2^{**}, \ldots, x_8^{**})^T$ given in Table 3.8, which is a solution that is beyond the scope of the Newton-Kantorovich theorem. Observe that $\|\mathbf{x}^{**}\| = 2.0087\ldots > \rho_2 = 1.4070\ldots$

In Table 3.9 we show the errors $\|\mathbf{x}^{**} - \tilde{\mathbf{x}}_n\|$, the error estimates and the sequence $\{\|\mathbb{F}(\tilde{\mathbf{x}}_n)\|\}$. From the last, we notice that the vector shown in Table 3.8 is a good approximation of the solution of system (1.52)-(3.7) with $m = 8$.

Finally, by interpolating the values of Table 3.8 and taking into account the boundary conditions, we obtain the solution denoted by $\tilde{\tilde{x}}$ and drawn in Figure 3.2.

n	$\|\mathbf{x}^{**} - \tilde{\mathbf{x}}_n\|$	$\|t^* - t_n\|$	$\|\mathbb{F}(\tilde{\mathbf{x}}_n)\|$
0	$6.5292\ldots \times 10^{-2}$	$6.7838\ldots \times 10^{-2}$	$1.4652\ldots \times 10^{-2}$
1	$3.4774\ldots \times 10^{-3}$	$6.0238\ldots \times 10^{-3}$	$5.2164\ldots \times 10^{-4}$
2	$9.8353\ldots \times 10^{-6}$	$5.8199\ldots \times 10^{-5}$	$1.5985\ldots \times 10^{-6}$
3	$8.0245\ldots \times 10^{-11}$	$5.5524\ldots \times 10^{-9}$	$1.2841\ldots \times 10^{-11}$

Table 3.9: Absolute errors, error estimates and $\{\|\mathbb{F}(\tilde{\mathbf{x}}_n)\|\}$

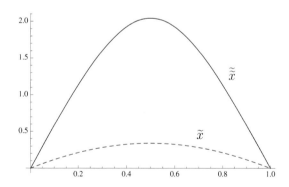

Figure 3.2: Approximated solutions of (1.44)-(1.45) with $\phi(x(s))$ defined in (3.7)

3.3 Operators with ω-Lipschitz k-th-derivative

From the semilocal convergence result given by Huang for Newton's method, the Huang theorem (Theorem 2.3), we observe that we can generalize the convergence conditions required in that result for sufficiently Fréchet differentiable operators. So, we suppose that the operator F satisfies the following condition:

$$\|F^{(k)}(x) - F^{(k)}(y)\| \le \omega(\|x - y\|), \quad x, y \in \Omega, \tag{3.10}$$

where $\omega : [0, +\infty) \longrightarrow \mathbb{R}$ is a nondecreasing continuos function such that $\omega(0) = 0$ and $k \ge 2$.

Notice that we also generalize the semilocal convergence result given in Section 2.1, where the condition required to the operator F is (3.10) with $k = 2$. For this, we have modified the technique of proving the semilocal convergence of Newton's method that is used there.

Note that the corresponding semilocal convergence results obtained from the generalizations mentioned previously are particular cases of the one presented in this section. Indeed, from (3.10), we have

$$\|F^{(k)}(x)\| - \|F^{(k)}(x_0)\| \le \|F^{(k)}(x) - F^{(k)}(x_0)\| \le \omega(\|x - x_0\|) \le \omega(\|x\| + \|x_0\|), \quad x \in \Omega,$$

so that we can consider

$$\|F^{(k)}(x)\| \le \|F^{(k)}(x_0)\| + \omega(\|x\| + \|x_0\|) = \tilde{\omega}(\|x\|),$$

where $\tilde{\omega} : [0, +\infty) \longrightarrow \mathbb{R}$ is a continuous monotonous function such that $\tilde{\omega}(0) \ge 0$.

In addition, we see that the domain of starting points for Newton's method is modified, but not reduced, with regard to those obtained from the Kantorovich and Huang results.

Section 3.3 is organized as follows. We begin in Section 3.3.1 by recalling the concept of majorizing sequence and give a general semilocal convergence result. In Section 3.3.2, we prove the existence of a majorant function. Then, in Section 3.3.3, we present, under the new convergence conditions introduced previously, a semilocal convergence result for Newton's method that includes information about the existence of solution. In Section 3.3.4, we give results on the uniqueness of solution and a priori error estimates that leads to the quadratic convergence of Newton's method. In Section 3.3.5, an application, where a boundary value problem is involved, is included that shows the advantages of the analysis of the semilocal convergence of Newton's method given previously with respect to the Newton-Kantorovich theorem. Finally, in Section 3.3.6, we relax the semilocal convergence conditions used in the study presented in the previous sections.

3.3.1   Existence of a majorizing sequence

Kantorovich and Huang prove the semilocal convergence of Newton's method in the Banach space X from construction of majorizing sequences. Once again, the procedure usually used to construct a majorizing real sequence for Newton's method in X is to find a real function $f : I \subseteq \mathbb{R} \longrightarrow \mathbb{R}$, consider the real sequence defined in (4.1) from the majorant function and determine the real interval I.

In the case analyzed by Huang, Section 2.1.1, we consider that $f(t)$ is a cubic polynomial and, as F'' is Lipschitz continuous in Ω, if

$$f''(t) - f''(t_0) = K(t - t_0), \quad t - t_0 \ge 0, \quad -\frac{1}{f'(t_0)} = \beta, \quad -\frac{f(t_0)}{f'(t_0)} = \eta, \quad f''(t_0) = \delta,$$

and solve the corresponding problem of interpolation fitting with $t_0 = 0$, we obtain polynomial (2.2), $\phi(t) = \frac{K}{6}t^3 + \frac{\delta}{2}t^2 - \frac{t}{\beta} + \frac{\eta}{\beta}$.

There are two important features to consider when looking for a real function that provides a majorizing real sequence: the real function depends fundamentally on the condition imposed on the operator F (condition (B2) in Huang's study) and the conditions imposed on the starting point x_0 that fix the initial parameters (condition (B1) in Huang's study).

The main aim of this section is to obtain a result that allows guaranteeing the existence of a majorizing real sequence for Newton's method under condition (3.10), with $k \geq 2$, for the operator F and the corresponding conditions for the starting point x_0 of Newton's method. For this, we use a majorant function, in the sense of Definition 1.26, and construct a majorizing real sequence. These kinds of results are usually referred to as general semilocal convergence theorems. In next section, from solving an initial value problem, we obtain a majorant function and semilocal convergence results for Newton's method.

So, following the ideas shown in Section 2.1, where condition (3.10) is considered with $k = 2$ and later relaxed, when the condition is centered in x_0, we consider that $F : \Omega \subseteq X \longrightarrow Y$ is a k-times continuously Fréchet differentiable operator and suppose that there exists $f \in \mathcal{C}^j([t_0, +\infty))$, with $t_0 \in \mathbb{R}$ and $j \geq k$, that satisfies:

(L1) There exists the operator $\Gamma_0 = [F'(x_0)]^{-1} \in \mathcal{L}(Y, X)$, for some $x_0 \in \Omega$, with $\|\Gamma_0\| \leq -\frac{1}{f'(t_0)}$ and $\|\Gamma_0 F(x_0)\| \leq -\frac{f(t_0)}{f'(t_0)}$; moreover, $\|F^{(i)}(x_0)\| \leq f^{(i)}(t_0)$ with $i = 2, 3, \ldots, k$ and $k \geq 2$,

(L2) $\|F^{(k)}(x) - F^{(k)}(x_0)\| \leq f^{(k)}(t) - f^{(k)}(t_0)$, for $\|x - x_0\| \leq t - t_0$, $x \in \Omega$ and $t \in [t_0, +\infty)$.

In addition, we have to prove that the corresponding real sequence given by (4.1) is nondecreasing and convergent to a solution of the equation $f(t) = 0$, so that the real sequence $\{t_n\}$ majorizes the sequence $\{x_n\}$ given by Newton's method in X, and the convergence of $\{x_n\}$ is then guaranteed.

We begin studying the existence of solution of the equation $f(t) = 0$, what is essential so that the sequence $\{t_n\}$ is convergent.

Theorem 3.23. *Suppose that there exist $f \in \mathcal{C}^j([t_0, +\infty))$, with $j \geq k$ and $t_0 \in \mathbb{R}$, such that (L1)-(L2) are satisfied.*

(a) *If there exists a solution $\alpha \in (t_0, +\infty)$ of the equation $f'(t) = 0$, then α is the unique minimum of f in $[t_0, +\infty)$ and f is nonincreasing in $[t_0, \alpha)$.*

(b) *If $f(\alpha) \leq 0$, then the equation $f(t) = 0$ has at least one solution in $[t_0, \alpha)$. Moreover, if t^* is the smallest solution of $f(t) = 0$ in $[t_0, \alpha)$, we have $t_0 < t^* \leq \alpha$.*

Proof. First, as $f^{(k)}(t_0) \geq 0$ follows from (L1) and (L2) is satisfied, then $f^{(k)}(t) \geq 0$ in $[t_0, +\infty)$ and $f^{(k-1)}(t)$ is nondecreasing in $[t_0, +\infty)$. Besides, $f^{(k-1)}(t) \geq 0$, since $f^{(k-1)}(t_0) \geq 0$. Repeating the previous reasoning until the second derivative, we obtain $f''(t) \geq 0$ in $[t_0, +\infty)$. Hence, f is convex in $[t_0, +\infty)$, α is the unique minimum of f in $[t_0, +\infty)$ and f' is nondecreasing in $[t_0, +\infty)$.

Second, as $f'(t_0) < 0$, $f'(\alpha) = 0$ and f' is nondecreasing in $[t_0, +\infty)$, then $f'(t) \leq 0$ in $[t_0, \alpha)$, so that f is nonincreasing in $[t_0, \alpha)$.

Third, if $f(\alpha) < 0$, as $f(t_0) > 0$ and f is continuous, then f has at least one zero t^* in (t_0, α). As f is nonincreasing in $[t_0, \alpha)$, then t^* is the unique zero of f in (t_0, α).

Finally, if $f(\alpha) = 0$, then α is a double zero of f and $t^* = \alpha$. ∎

Once we have established the conditions so that the equation $f(t) = 0$ has a solution, we establish the convergence of the majorizing real sequence $\{t_n\}$ given by (4.1) in the next theorem. Notice that if there exists the function f, then it has the same properties as Kantorovich's polynomial (1.23), since $f(t) > 0$, $f'(t) < 0$ and $f''(t) > 0$ in (t_0, α), so that the convergence of the real sequence given in (4.1) is then guaranteed in a way analogous to that of sequence (1.22) and such as we have seen in Theorem 1.21.

Theorem 3.24. *Suppose that there exist $f \in \mathcal{C}^j([t_0, +\infty)$, with $j \geq k$ and $t_0 \in \mathbb{R}$, such that (L1)-(L2) are satisfied. If there exists a solution $\alpha \in (t_0, +\infty)$ of $f'(t) = 0$ such that $f(\alpha) \leq 0$, then the scalar sequence $\{t_n\}$, given by (4.1), is nondecreasing and converges to the smallest positive solution t^* of the equation $f(t) = 0$.*

The following is to prove that the real sequence $\{t_n\}$, given by (4.1), majorizes Newton's sequence $\{x_n\}$ in X.

Theorem 3.25. *Suppose that there exist $f \in \mathcal{C}^j([t_0, +\infty))$, with $j \geq k$ and $t_0 \in \mathbb{R}$, such that (L1)-(L2) are satisfied. If there exists a solution $\alpha \in (t_0, +\infty)$ of $f'(t) = 0$ such that $f(\alpha) \leq 0$ and $x_n \in \Omega$, for all $n \in \mathbb{N}$, then the sequence $\{t_n\}$, given by (4.1), majorizes Newton's sequence $\{x_n\}$; i.e.:*

$$\|x_n - x_{n-1}\| \leq t_n - t_{n-1}, \quad \text{for all} \quad n \in \mathbb{N}.$$

Proof. First of all, from (L1), we have

$$\|x_1 - x_0\| = \|\Gamma_0 F(x_0)\| \leq -\frac{f(t_0)}{f'(t_0)} = t_1 - t_0.$$

After that, we prove the thesis of the theorem from the next four recurrence relations (for $n \in \mathbb{N}$):

(i_n) There exists $\Gamma_n = [F'(x_n)]^{-1}$ and $\|\Gamma_n\| \leq -\dfrac{1}{f'(t_n)}$,

(ii_n) $\|F^{(i)}(x_n)\| \leq f^{(i)}(t_n)$, $i = 2, 3, \ldots, k$,

(iii_n) $\|F(x_n)\| \leq f(t_n)$,

(iv_n) $\|x_{n+1} - x_n\| \leq t_{n+1} - t_n$.

We prove items (i_n)-(ii_n)-(iii_n)-(iv_n) by mathematical induction on n. The case (i_1)-(iii_1)-(iii_1)-(iv_1) is analogous to which we do for the inductive step, no more to consider conditions (L1)-(L2). For the inductive step, we suppose that (i_n)-(ii_n)-(iii_n)-(iv_n) are true for $n = 1, 2, \ldots, d - 1$ and see that (i_d)-(ii_d)-(iii_d)-(iv_d) are true.

To prove (i_d), we consider Taylor's series of F for

$$I - \Gamma_{d-1}F'(x_d) = \Gamma_{d-1}\left(F'(x_{d-1}) - F'(x_d)\right)$$

$$= \Gamma_{d-1}\left(F'(x_{d-1}) - \sum_{j=1}^{k-1}\frac{1}{(j-1)!}F^{(j)}(x_{d-1})(x_d - x_{d-1})^{j-1}\right.$$

$$\left. - \frac{1}{(k-2)!}\int_{x_{d-1}}^{x_d} F^{(k)}(z)(x_d - z)^{k-2}\,dz\right)$$

$$= -\Gamma_{d-1}\left(\sum_{j=2}^{k-1}\frac{1}{(j-1)!}F^{(j)}(x_{d-1})(x_d - x_{d-1})^{j-1}\right.$$

$$\left. - \frac{1}{(k-2)!}\int_{x_{d-1}}^{x_d} (F^{(k)}(z) \pm (F^{(k)}(x_0)))(x_d - z)^{k-2}\,dz\right)$$

$$= -\Gamma_{d-1}\left(\sum_{j=2}^{k-1}\frac{1}{(j-1)!}F^{(j)}(x_{d-1})(x_d - x_{d-1})^{j-1}\right.$$

$$+ \frac{1}{(k-2)!}\int_{x_{d-1}}^{x_d} F^{(k)}(x_0)(x_d - z)^{k-2}\,dz$$

$$\left. + \frac{1}{(k-2)!}\int_{x_{d-1}}^{x_d} (F^{(k)}(z) - (F^{(k)}(x_0)))(x_d - z)^{k-2}\,dz\right).$$

Moreover, if $z \in [x_{d-1}, x_d]$ and $s \in [t_{d-1}, t_d]$, then $z = x_{d-1} + \tau(x_d - x_{d-1})$ and $s = t_{d-1} + \tau(t_d - t_{d-1})$ with $\tau \in [0,1]$, so that

$$\|z - x_0\| \le \tau\|x_d - x_{d-1}\| + \|x_{d-1} - x_{d-2}\| + \cdots + \|x_1 - x_0\|$$

$$\le \tau(t_d - t_{d-1}) + t_{d-1} - t_{d-2} + \cdots + t_1 - t_0$$

$$= s - t_0.$$

As a consequence,

$$\|I - \Gamma_{d-1}F'(x_d)\| \leq \|\Gamma_{d-1}\| \sum_{j=2}^{k-1} \frac{1}{(j-1)!} \|F^{(j)}(x_{d-1})\| \|x_d - x_{d-1}\|^{j-1}$$

$$+ \frac{1}{(k-2)!} \|\Gamma_{d-1}\| \|F^{(k)}(x_0)\| \|x_d - x_{d-1}\|^{k-1} \left(\int_0^1 (1-\tau)^{k-2} d\tau\right)$$

$$+ \frac{1}{(k-2)!} \|\Gamma_{d-1}\| \int_0^1 \|F^{(k)}(x_{d-1} + \tau(x_d - x_{d-1})) - F^{(k)}(x_0)\|$$

$$\times (1-\tau)^{k-2} d\tau \|x_d - x_{d-1}\|^{k-1}$$

$$\leq -\frac{1}{f'(t_{d-1})} \sum_{j=2}^{k-1} \frac{f^{(j)}(t_{d-1})}{(j-1)!} (t_d - t_{d-1})^{j-1}$$

$$- \frac{1}{(k-1)!} \frac{f^{(k)}(t_0)}{f'(t_{d-1})} (t_d - t_{d-1})^{k-1}$$

$$- \frac{1}{(k-2)!} \frac{1}{f'(t_{d-1})} \int_0^1 (f^{(k)}(t_{d-1} + \tau(t_d - t_{d-1})) - (F^{(k)}(t_0))$$

$$\times (1-\tau)^{k-2} d\tau (t_d - t_{d-1})^{k-1}$$

$$= -\frac{1}{f'(t_{d-1})} \left(\sum_{j=2}^{k-1} \frac{f^{(j)}(t_{d-1})}{(j-1)!} (t_d - t_{d-1})^{i-1}\right.$$

$$\left.+ \frac{1}{(k-2)!} \int_0^1 f^{(k)}(t_{d-1} + \tau(t_d - t_{d-1}))(1-\tau)^{k-2} d\tau (t_d - t_{d-1})^{k-1}\right)$$

$$= 1 - \frac{1}{f'(t_{d-1})} \left(\sum_{j=1}^{k-1} \frac{f^{(j)}(t_{d-1})}{(j-1)!} (t_d - t_{d-1})^{j-1}\right.$$

$$\left.+ \frac{1}{(k-2)!} \int_{t_{d-1}}^{t_d} f^{(k)}(u)(t_d - u)^{k-2} du\right)$$

$$= 1 - \frac{f'(t_d)}{f'(t_{d-1})}$$

$$< 1,$$

since $t_{d-1} < t_d \leq t^*$, $f''(t) > 0$ in $[t_0, +\infty)$ and $\frac{f'(t_d)}{f'(t_{d-1})} \in (0,1)$. In addition, by the Banach lemma on invertible operators, it follows that there exists the operator Γ_d and $\|\Gamma_d\| \leq -\frac{1}{f'(t_d)}$.

To prove (ii_d), we distinguish two cases: $i \in \{2, 3, \ldots, k-1\}$ and $i = k$. If $i \in \{2, 3, \ldots, k-$

1}, we see, from Taylor's series, that

$$F^{(i)}(x_d) = \sum_{j=1}^{k-i} \frac{1}{(j-1)!} F^{(j+i-1)}(x_{d-1})(x_d - x_{d-1})^{j-1}$$

$$+ \frac{1}{(k-i-1)!} \int_{x_{d-1}}^{x_d} F^{(k)}(z)(x_d - z)^{k-i-1} \, dz$$

$$= \sum_{j=1}^{k-i} \frac{1}{(j-1)!} F^{(j+i-1)}(x_{d-1})(x_d - x_{d-1})^{j-1}$$

$$+ \frac{1}{(k-i-1)!} \left(\int_{x_{d-1}}^{x_d} F^{(k)}(x_0)(x_d - z)^{k-i-1} \, dz \right.$$

$$\left. + \int_{x_{d-1}}^{x_d} (F^{(k)}(z) - F^{(k)}(x_0))(x_d - z)^{k-i-1} \, dz \right).$$

In addition, if $\|z - x_0\| \le s - t_0$, where $z \in [x_{d-1}, x_d]$ and $s \in [t_{d-1}, t_d]$, and $\|x_d - x_{d-1}\| \le t_d - t_{d-1}$, then

$$\|F^{(i)}(x_d)\| \le \sum_{j=1}^{k-i} \frac{1}{(j-1)!} \|F^{(j+i-1)}(x_{d-1})\| \|x_d - x_{d-1}\|^{j-1}$$

$$+ \frac{1}{(k-i-1)!} \left(\int_0^1 \|F^{(k)}(x_0)\|(1-\tau)^{k-i-1} d\tau \, \|x_d - x_{d-1}\|^{k-i} \right.$$

$$\left. + \int_0^1 \|F^{(k)}(x_{d-1} + \tau(x_d - x_{d-1})) - F^{(k)}(x_0)\|(1-\tau)^{k-i-1} d\tau \, \|x_d - x_{d-1}\|^{k-i} \right)$$

$$\le \sum_{j=1}^{k-i} \frac{f^{(j+i-1)}(t_{d-1})}{(j-1)!}(t_d - t_{d-1})^{j-1}$$

$$+ \frac{1}{(k-i-1)!} \int_0^1 f^{(k)}(t_{d-1} + \tau(t_d - t_{d-1}))(1-\tau)^{k-i-1} d\tau \, (t_d - t_{d-1})^{k-i}$$

$$= f^{(i)}(t_d)$$

Now, if $i = k$, then

$$\|F^{(k)}(x_d)\| = \|F^{(k)}(x_d) - F^{(k)}(x_0) + F^{(k)}(x_0)\|$$

$$\le \|F^{(k)}(x_d) - F^{(k)}(x_0)\| + \|F^{(k)}(x_0)\|$$

$$\le f^{(k)}(t_d) - f^{(k)}(t_0) + f^{(k)}(t_0)$$

$$= f^{(k)}(t_d).$$

The proof of item (iii_d) is completely analogous to that of the first case of item (ii_d) and item (iv_d) follows immediately, since

$$\|x_{d+1} - x_d\| \le \|\Gamma_d\| \|F(x_d)\| \le -\frac{f(t_d)}{f'(t_d)} = t_{d+1} - t_d.$$

Finally, as a consequence, items (i_n)-(ii_n)-(iii_n)-(iv_n) are true for all positive integers n by mathematical induction. ∎

Once we have proved that $\{t_n\}$ is a majorizing real sequence of the sequence $\{x_n\}$, we are ready to guarantee the semilocal convergence of Newton's sequence in the Banach space X.

Theorem 3.26. (General semilocal convergence) *Let $F : \Omega \subseteq X \longrightarrow Y$ be a k ($k \geq 2$) times continuously Fréchet differentiable operator defined on a non-empty open convex domain Ω of a Banach space X with values in a Banach space Y. Suppose that there exist $f \in C^j([t_0, +\infty))$, with $j \geq k$ and $t_0 \in \mathbb{R}$, such that (L1)-(L2) are satisfied and a root $\alpha \in (t_0, +\infty)$ of $f'(t) = 0$ such that $f(\alpha) \leq 0$, and $B(x_0, t^* - t_0) \subset \Omega$. Then, the sequence $\{x_n\}$ given by Newton's method, converges to a solution x^* of $F(x) = 0$ starting at x_0. Moreover, $x_n, x^* \in \overline{B(x_0, t^* - t_0)}$ and*

$$\|x^* - x_n\| \leq t^* - t_n, \quad for\ all\quad n = 0, 1, 2, \ldots,$$

where $\{t_n\}$ is defined in (4.1).

Proof. From (L1)-(L2), it is clear that x_1 is well-defined and $\|x_1 - x_0\| \leq t_1 - t_0 < t^* - t_0$, so that $x_1 \in B(x_0, t^* - t_0) \subset \Omega$.

We now suppose that x_n is well-defined and $x_n \in B(x_0, t^* - t_0)$ for $n = 1, 2, \ldots, d - 1$. After that, by the last lemma, the operator Γ_{d-1} exists and $\|\Gamma_{d-1}\| \leq -\dfrac{1}{f'(t_{d-1})}$. In addition, x_d is well-defined. Moreover, since $\|x_{n+1} - x_n\| \leq t_{n+1} - t_n$ for $n = 1, 2, \ldots, d - 1$, we have

$$\|x_d - x_0\| \leq \sum_{i=0}^{d-1} \|x_{i+1} - x_i\| \leq \sum_{i=0}^{d-1}(t_{i+1} - t_i) = t_d - t_0 < t^* - t_0,$$

so that $x_d \in B(x_0, t^* - t_0)$. Therefore, Newton's sequence $\{x_n\}$ is well-defined and $x_n \in B(x_0, t^* - t_0)$ for all $n \geq 0$.

By the last lemma, we also have that $\|F(x_n)\| \leq f(t_n)$ and

$$\|x_{n+1} - x_n\| \leq \|\Gamma_n\|\|F(x_n)\| \leq -\frac{f(t_n)}{f'(t_n)} = t_{n+1} - t_n,$$

for all $n \geq 0$, so that $\{t_n\}$ is a majorizing real sequence of $\{x_n\}$ and is convergent. Moreover, as $\lim_n t_n = t^*$ if $x^* = \lim_n x_n$, then $\|x^* - x_n\| \leq t^* - t_n$, for all $n = 0, 1, 2, \ldots$ Furthermore, as $\|F(x_n)\| \leq f(t_n)$, for all $n = 0, 1, 2, \ldots$, then, by letting $n \to +\infty$, it follows $F(x^*) = 0$ by the continuity of F. ∎

3.3.2   Existence of a majorant function

The next aim is to obtain a semilocal convergence result for Newton's method from the general semilocal convergence theorem, Theorem 3.26, established in the last section. This result generalizes Huang's result as well as other results known until now that also generalize Huang's result. To apply Theorem 3.26, we have to prove the existence of a majorant function that satisfies the corresponding conditions imposed in the theorem. Moreover, the operator F must satisfy condition (3.10). Observe that condition (3.10) is reduced to condition (B2) of Huang if $k = 2$ and $\omega(z) = Kz$. In addition, condition (3.10) is also reduced to

$$\|F^{(k)}(x) - F^{(k)}(y)\| \leq K\|x - y\|^p, \quad K \geq 0, \quad x, y \in \Omega, \quad p \in [0, 1], \quad k \geq 2,$$

if $\omega(z) = Kz^p$, so that condition (3.10) also generalizes the case in which $F^{(k)}$ is Hölder continuous in Ω.

For constructing a majorant function, we suppose first the following conditions:

(O1) There exists $\Gamma_0 = [F'(x_0)]^{-1} \in \mathcal{L}(Y, X)$, for some $x_0 \in \Omega$, with $\|\Gamma_0\| \leq \beta$ and $\|\Gamma_0 F(x_0)\| \leq \eta$; moreover, $\|F^{(i)}(x_0)\| \leq b_i$, with $i = 2, 3, \ldots, k$ and $k \geq 2$.

(O2) There exists a continuous and nondecreasing function $\omega : [0, +\infty) \longrightarrow \mathbb{R}$ such that $\|F^{(k)}(x) - F^{(k)}(y)\| \leq \omega(\|x - y\|)$ for $x, y \in \Omega$ and $\omega(0) = 0$.

After that, we look for a scalar function f that majorizes the operator F. As we have seen above, this function depends on the starting point x_0, so that, once $x_0 \in \Omega$ is fixed, we have

$$\|F^{(k)}(x) - F^{(k)}(x_0)\| \leq \omega(\|x - x_0\|) \leq \omega(t - t_0) = f^{(k)}(t) - f^{(k)}(t_0) \quad \text{for} \quad \|x - x_0\| \leq t - t_0.$$

Observe that we cannot apply the reasoning used in the Kantorovich and Huang cases, since we cannot think in a elemental function f that satisfies the condition

$$f^{(k)}(t) - f^{(k)}(t_0) = \omega(t - t_0), \quad t \in [t_0, +\infty), \tag{3.11}$$

for any function ω. Thus, we cannot formulate this problem as a problem of interpolation fitting. In this case, we proceed differently by considering the following initial value problem:

$$\begin{cases} y^{(k)}(t) = b_k + \omega(t - t_0), \\ y(t_0) = \dfrac{\eta}{\beta}, \quad y'(t_0) = -\dfrac{1}{\beta}, \\ y''(t_0) = b_2, \quad y'''(t_0) = b_3, \quad \ldots, \quad y^{(k-1)}(t_0) = b_{k-1}, \end{cases}$$

whose solution $f(t)$, if it exists, satisfies conditions (L1)-(L2) of the general semilocal convergence theorem, Theorem 3.26, and condition (3.11), since we have

$$\|F^{(k)}(x) - F^{(k)}(x_0)\| \leq \omega(\|x - x_0\|) \leq \omega(t - t_0) = f^{(k)}(t) - b_k = f^{(k)}(t) - f^{(k)}(t_0),$$

for $x, x_0 \in \Omega$ and $t \in [t_0, +\infty)$ such that $\|x - x_0\| \leq t - t_0$, so that condition (L2) of Theorem 3.26 is satisfied. Therefore, if the initial value problem has a solution $f(t)$, we have proved that a real function $f(t)$ exists, satisfying conditions (L1)-(L2) exists. In addition, it is clear that $f^{(k)}(t) \geq 0$ for $t \in [t_0, +\infty)$, since $\omega(t - t_0) \geq 0$ for $t \in [t_0, +\infty)$.

Theorem 3.27. *Suppose that the function $\omega(t - t_0)$ is continuous and $\omega(t - t_0) \geq 0$ in $[t_0, +\infty)$. For any nonnegative real positive numbers $\beta \neq 0$, η, $b_2, b_3, \ldots, b_k$, there exists only one solution $f(t)$ of the last initial value problem in $[t_0, +\infty)$; that is:*

$$f(t) = \int_{t_0}^{t} \int_{t_0}^{\theta_{k-1}} \cdots \int_{t_0}^{\theta_1} \omega(\xi - t_0) \, d\xi \, d\theta_1 \cdots d\theta_{k-1}$$

$$+ \frac{b_k}{k!}(t - t_0)^k + \cdots + \frac{b_2}{2!}(t - t_0)^2 - \frac{t - t_0}{\beta} + \frac{\eta}{\beta}. \tag{3.12}$$

Notice that if $k = 2$ and $\omega(z) = Kz$, the operator F'' is Lipschitz continuous in Ω and function (3.12) is reduced to Huang's polynomial (2.2). Moreover, if $F^{(k)}$ with $k \geq 2$ is Lipschitz continuous in Ω with Lipschitz constant λ, then function (3.12) is the following polynomial of degree $k + 1$:

$$\Psi(t) = \frac{\lambda}{(k+1)!}(t - t_0)^{k+1} + \frac{b_k}{k!}(t - t_0)^k + \cdots + \frac{b_2}{2!}(t - t_0)^2 - \frac{t - t_0}{\beta} + \frac{\eta}{\beta},$$

allowing us to define majorizing sequences for Newton's method in the Banach space X.

In addition, function (3.12) has the properties given in the next result, whose proof is similar to that given for Theorem 2.15.

Theorem 3.28. *Let f and ω be the functions defined respectively in (3.12) and (3.10).*

(a) *There exists only one positive solution $\alpha > t_0$ of the equation $f'(t) = 0$, which is the unique minimum of $f(t)$ in $[t_0, +\infty)$ and $f(t)$ is nonincreasing in $[t_0, \alpha)$.*

(b) *If $f(\alpha) \leq 0$, then the equation $f(t) = 0$ has at least one solution in $[t_0, +\infty)$. Moreover, if t^* is the smallest solution of $f(t) = 0$ in $[t_0, +\infty)$, we have $t_0 < t^* \leq \alpha$.*

3.3.3   Semilocal convergence

Now, we are ready to establish a semilocal convergence result for Newton's method when the operator F satisfies condition (3.10). For this, we suppose conditions (O1)-(O2) and the following one:

(O3) $f(\alpha) \leq 0$, where f is the function defined in (3.12) and α is the unique positive solution of $f'(t) = 0$, and $B(x_0, t^* - t_0) \subset \Omega$, where t^* is the smallest positive solution of $f(t) = 0$ in $[t_0, +\infty)$.

Then, we present the next theorem, whose proof follows immediately from Theorem 3.26.

Theorem 3.29. *Let $F : \Omega \subseteq X \longrightarrow Y$ be a k ($k \geq 2$) times continuously Fréchet differentiable operator defined on a non-empty open convex domain Ω of a Banach space X with values in a Banach space Y and $f(t)$ the function defined in (3.12). Suppose that conditions (O1)-(O2)-(O3) are satisfied. Then, the sequence $\{x_n\}$, given by Newton's method, converges to a solution x^* of $F(x) = 0$ starting at x_0. Moreover, $x_n, x^* \in \overline{B(x_0, t^* - t_0)}$ and*

$$\|x^* - x_n\| \leq t^* - t_n, \quad for \ all \quad n = 0, 1, 2, \dots,$$

where $\{t_n\}$ is defined in (4.1).

3.3.4   Uniqueness of solution and order of convergence

After proving the semilocal convergence of Newton's method and locating the solution x^*, we prove the uniqueness of x^*. Before, we give the following technical lemma that is used later.

Lemma 3.30. *Under the conditions of Theorem 3.26, if the scalar function $f(t)$ given in (3.12) has two real zeros t^* and t^{**} such that $t_0 < t^* \leq t^{**}$ and $x \in \overline{B(x_0, t^{**} - t_0)} \cap \Omega$, then*

$$\|F''(x)\| \leq f''(t), \quad for \quad \|x - x_0\| \leq t - t_0.$$

Proof. To prove the theorem, we distinguish the case $k = 2$ of the case $k \geq 3$. The case $k = 2$ follows as in Section 2.1.3 and the case $k \geq 3$ follows as in proof of Lemma 3.19, without more that take into account the following mathematical decomposition:

$$F''(x) = \sum_{i=2}^{k-1} \frac{1}{(i-2)!} F^{(i)}(x_0)(x-x_0)^{i-2} + \frac{1}{(k-3)!} \int_{x_0}^{x} F^{(k)}(z)(x-z)^{k-3} \, dz$$

$$= \sum_{i=2}^{k-1} \frac{1}{(i-2)!} F^{(i)}(x_0)(x-x_0)^{i-2}$$

$$+ \frac{1}{(k-3)!} \int_{x_0}^{x} \left(F^{(k)}(z) \pm F^{(k)}(x_0) \right) (x-z)^{k-3} \, dz$$

$$= \sum_{i=2}^{k} \frac{1}{(i-2)!} F^{(i)}(x_0)(x-x_0)^{i-2}$$

$$+ \frac{1}{(k-3)!} \int_{0}^{1} \left(F^{(k)}\left(x_0 + \tau(x-x_0)\right) - F^{(k)}(x_0) \right) (x-x_0)^{k-2}(1-\tau)^{k-3} \, d\tau,$$

from Taylor's series.   ∎

Note that if $f(t)$ has two real zeros t^* and t^{**} such that $t_0 < t^* \leq t^{**}$, then the uniqueness of the solution follows from the next theorem, whose proof follows similarly to that given for Theorem 3.6.

Theorem 3.31. *Under the hypotheses of Theorem 3.26, if the scalar function $f(t)$ has two real zeros t^* and t^{**} such that $t_0 < t^* \leq t^{**}$, then the solution x^* is unique in $B(x_0, t^{**}-t_0) \cap \Omega$ if $t^* < t^{**}$ or in $\overline{B(x_0, t^* - t_0)}$ if $t^{**} = t^*$.*

Observe that the last theorem is a generalization of that obtained under Kantorovich's conditions.

Next, we see the quadratic convergence of Newton's method under conditions (L1)-(L2). Notice first that if $f(t)$ has two real zeros t^* and t^{**} such that $t_0 < t^* \leq t^{**}$, we can then write

$$f(t) = (t^* - t)(t^{**} - t)g(t)$$

with $g(t^*) \neq 0$ and $g(t^{**}) \neq 0$. Next, from Ostrowski's technique [65], we give a theorem which provides some a priori error estimates for Newton's method. The proof of the theorem is analogous to that of Theorem 2.18.

Theorem 3.32. *Under the hypotheses of Theorem 3.26 and assuming that $f(t) = (t^* - t)(t^{**} - t)g(t)$, where $g(t)$ is a function such that $g(t^*) \neq 0$ and $g(t^{**}) \neq 0$, we have the following:*

(a) *If $t^* < t^{**}$ and there exist $m_1 > 0$ and $M_1 > 0$ such that $m_1 \leq \min\{Q_1(t); t \in [t_0, t^*]\}$ and $M_1 \geq \max\{Q_1(t); t \in [t_0, t^*]\}$, where $Q_1(t) = \frac{(t^{**}-t)g'(t)-g(t)}{(t^*-t)g'(t)-g(t)}$, then*

$$\frac{(t^{**} - t^*)\theta^{2^n}}{m_1 - \theta^{2^n}} \leq t^* - t_n \leq \frac{(t^{**} - t^*)\Delta^{2^n}}{M_1 - \Delta^{2^n}}, \quad n \geq 0,$$

where $\theta = \frac{t^*}{t^{**}}m_1$, $\Delta = \frac{t^*}{t^{**}}M_1$ and provided that $\theta < 1$ and $\Delta < 1$.

(b) *If $t^* = t^{**}$ and there exist $m_2 > 0$ and $M_2 > 0$ such that $m_2 \leq \min\{Q_2(t); t \in [t_0, t^*]\}$ and $M_2 \geq \max\{Q_2(t); t \in [t_0, t^*]\}$, where $Q_2(t) = \frac{(t^*-t)g'(t)-g(t)}{(t^*-t)g'(t)-2g(t)}$, then*

$$m_2^n t^* \leq t^* - t_n \leq M_2^n t^*, \quad n \geq 0,$$

provided that $m_2 < 1$ and $M_2 < 1$.

From the last theorem, it follows that the convergence of Newton's method, under conditions (L1)-(L2), is quadratic if $t^* < t^{**}$ and linear if $t^* = t^{**}$.

3.3.5   Application

We present a boundary value problem, where the previous study is applied. In particular, from the Theorem 3.29, we locate and separate solutions of the problem and guarantee the semilocal convergence of Newton's method to a solution of the problem.

In practice, we usually choose $t_0 = 0$, since function (3.12) has the same property of translation as Kantorovich's polynomial (see Remark 1.18) and majorizes the operator F, in the sense of Definition 1.26, independent of the value of t_0.

Example 3.33. A typical example of a boundary value problem is (1.47)-(1.48), that has been introduced in Section1.2.2,

$$\frac{d^2 x(s)}{ds^2} + \Psi(s, x(s)) = 0, \qquad x(a) = A, \qquad x(b) = B.$$

We have seen that we can then consider the second-order differential equation as a twice differentiable operator $\mathfrak{F} : \mathcal{C}^2([a, b]) \longrightarrow \mathcal{C}([a, b])$ such that

$$[\mathfrak{F}(x)](s) = \frac{d^2 x(s)}{ds^2} + \Psi(s, x(s)).$$

Then, the first two derivatives of $\mathfrak{F}$ are

$$\mathfrak{F}'(x) = \frac{d^2}{ds^2} + \Psi'_2(s, x(s))I, \qquad \mathfrak{F}''(x) = \Psi''_2(s, x(s))I_2,$$

where I_2 is the bilinear operator defined by $I_2 xy(s) = x(s)y(s)$.

Setting,

$$u_n(s) = x_{n+1}(s) - x_n(s),$$

$$v_n(s) = -\mathfrak{F}(x_n) = -\frac{d^2}{ds^2} x_n(s) - \Psi(s, x_n(s)), \quad n = 0, 1, 2, \ldots,$$

we arrive at the linear boundary value problems

$$\begin{cases} \dfrac{d^2 u_n}{ds^2} + \Psi'_2(s, x_n)u_n = \mathfrak{F}'(x_n)u_n = -\mathfrak{F}(x_n) = v_n(s), \\ u_n(a) = A, \quad u_n(b) = B, \quad n = 0, 1, 2, \ldots, \end{cases} \tag{3.13}$$

for the generation of Newton's sequence $\{x_n(s)\}$. It is assumed that $x_0(s)$ satisfies the boundary conditions.

Note that the solution of linear boundary value problems such as (3.13) for second-order differential equations is by no means trivial. One possible approach employs an integral equation technique, that we can see in [72].

For a specific example of a boundary value problem as the above, we consider

$$\frac{d^2 x}{ds^2} - sx(s)^6 + 1 = 0, \qquad x(0) = x(1) = 0. \tag{3.14}$$

Here the linear problems (3.13) are

$$\begin{cases} \dfrac{d^2 u_n}{ds^2} - 6sx_n^5 u_n = -\dfrac{d^2 x_n}{ds^2} + (sx_n^6 - 1), \\ u_n(0) = u_n(1) = 0, \quad n = 0, 1, 2, \ldots \end{cases} \tag{3.15}$$

Since a solution of (3.15) would vanish at the end points, a reasonable choice of the initial approximation, taking into account how the differential equation is, seems to be the parabola $x_0(s) = (2.9)s(1 - s)$, $s \in [0, 1]$. For this choice, (3.15) becomes

$$\begin{cases} \dfrac{d^2 u_0}{ds^2} - (1230.67)s^6(1 - s)^5 = (4.8) + (594.823)s^7(1 - s)^6, \\ u_0(0) = u_0(1) = 0. \end{cases}$$

The corresponding Fredholm integral equation (1.49) is

$$u_0(s) - \int_0^1 G(s, t)(1230.67)t^6(1 - t)^5 u_0(t)\, dt = \kappa(s), \tag{3.16}$$

where

$$\kappa(s) = (0.0115\ldots)s - (8.2614\ldots)s^9 + (39.6549\ldots)s^{10} - (81.1123\ldots)s^{11}$$
$$+ (90.1247\ldots)s^{12} - (57.1946\ldots)s^{13} + (19.6096\ldots)s^{14} - (2.8324\ldots)s^{15}.$$

Using the norm

$$\|x\| = \max\{\|x\|_\infty, \|x'\|_\infty, \|x''\|_\infty\}$$

for $\mathcal{C}^2([0, 1])$, where $\|x\|_\infty = \max_{[0,1]} |x(s)|$ is the norm for $\mathcal{C}([0, 1])$, we have that the linear integral operator $T : \mathcal{C}([0, 1]) \longrightarrow \mathcal{C}^2([0, 1])$ with kernel

$$T(s, t) = G(s, t)t^6(1 - t)^5, \quad 0 \le s, t \le 1,$$

has the bound $\|T\| \le 0.6288\ldots$, so that the integral equation (3.16) has a unique solution $u_0(s)$ which can be found by the Neumann series (see [68]), and

$$\|[\mathfrak{F}'(x_0)]^{-1}\mathfrak{F}(x_0)\| = \|u_0\| = \|(I - T)^{-1}\kappa\| \le \frac{\|\kappa\|}{1 - \|T\|} \le 0.0132\ldots,$$

by the Banach lemma on invertible operators. Moreover, as

$$u_0(s) = (I - T)^{-1} \int_0^1 G(s, t)[\mathfrak{F}(x_0)(x_0)](t)\, dt,$$

it follows

$$\|[\mathfrak{F}'(x_0)]^{-1}\| \le \frac{1}{1 - \|T\|} \le 2.6946\ldots$$

As a consequence, $\beta = 2.6946\ldots$ and $\eta = 0.0132\ldots$ Besides, since $\mathfrak{F}''(x) = -30sx^4 I_2$ is a bilinear operator from $\mathcal{C}^2([0, 1])$ into $\mathcal{C}([0, 1])$, we have $\|\mathfrak{F}''(x_0)\| \le 4.3814\ldots = b_2$. Analogously, we obtain $\|\mathfrak{F}'''(x_0)\| \le 24.5636\ldots = b_3$ and $\|\mathfrak{F}^{(iv)}(x_0)\| \le 104.6340\ldots = b_4$.

In addition, for Theorem 3.29 with $t_0 = 0$, we consider $\Omega = B(0,2) \subset \mathcal{C}^2([0,1])$, so that $\|\mathfrak{F}^{(iv)}(x) - \mathfrak{F}^{(iv)}(y)\| \leq 1440\|x - y\|$ and $\omega(t) = 1440t$. Therefore,

$$f(t) = (0.0049\ldots) - (0.3711\ldots)t + (2.1907\ldots)t^2 + (4.0939\ldots)t^3 + (64.3597\ldots)t^4,$$

which has two positive real zeros $t^* = 0.0145\ldots$ and $t^{**} = 0.0994\ldots$ Consequently, Theorem 3.29 guarantees the existence of a solution $x^*(s)$ of the nonlinear boundary value problem (3.14) and the convergence of Newton's sequence, $\{x_n(s)\}$, starting from $x_0(s) = (2.9)s(1-s) \in \Omega = B(0,2)$, to $x^*(s)$. Besides, from Theorem 3.29, we obtain that the domains of existence and uniqueness of solution are respectively

$$\{\nu \in \Omega : \|\nu(s) - x_0(s)\| \leq 0.0145\ldots\} \quad \text{and} \quad \{\nu \in \Omega : \|\nu(s) - x_0(s)\| < 0.0994\ldots\}.$$

Finally, note that we approximate later a solution of problem (3.14) in Example 3.36.

3.3.6   Relaxing convergence conditions

We have seen in Section 2.1 that condition (B2) of Huang can be relaxed by a center Lipschitz condition, a center Hölder condition and a center ω-Lipschitz condition at the point x_0 for the second derivative F'' of the operator F, so that the domain of starting points is modified, but not reduced, with respect to that given by Kantorovich. Moreover, we have also seen in case 3 of Section 2.1.3.6 that there are operators whose second derivative does not satisfy a Lipschitz condition in the entire domain Ω, so that condition (B2) is not satisfied, but they do at some point of the domain Ω. In addition, if we pay attention to the results obtained in the previous section and take into account what was said in Section 2.1.3, we observe that condition (3.10) is not necessary, since it is enough that the condition

$$\|F^{(k)}(x) - F^{(k)}(x_0)\| \leq \omega_0(\|x - x_0\|), \quad x \in \Omega,$$

is fulfilled where $\omega_0 : [0, +\infty) \longrightarrow \mathbb{R}$ is a nondecreasing continuous function such that $\omega_0(0) = 0$ and $k \geq 2$, instead of condition (3.10). According to the last, instead of Theorem 3.29, we establish the next semilocal convergence theorem for Newton's method under the following conditions:

(V1) There exists $\Gamma_0 = [F'(x_0)]^{-1} \in \mathcal{L}(Y, X)$, for some $x_0 \in \Omega$, with $\|\Gamma_0\| \leq \beta$ and $\|\Gamma_0 F(x_0)\| \leq \eta$; moreover, $\|F^{(i)}(x_0)\| \leq b_i$, with $i = 2, 3, \ldots, k$ and $k \geq 2$.

(V2) There exists a continuous and nondecreasing function $\omega_0 : [0, +\infty) \longrightarrow \mathbb{R}$ such that $\|F^{(k)}(x) - F^{(k)}(x_0)\| \leq \omega_0(\|x - x_0\|)$, for $x \in \Omega$, and $\omega_0(0) = 0$.

(V3) $f_0(\alpha) \leq 0$, where f_0 is the function defined in (3.12) with $\omega(t)$ replaced by $\omega_0(t)$ and α is the unique positive solution of $f_0'(t) = 0$, and $B(x_0, t^* - t_0) \subset \Omega$, where t^* is the smallest positive solution of $f_0(t) = 0$ in $[t_0, +\infty)$.

Note that, in this case, there is not a condition for the operator $F^{(k)}$ at various points of Ω, as it occurs in (O2), but only a center ω_0-Lipschitz condition required at the starting point x_0.

Theorem 3.34. *Let $F : \Omega \subseteq X \longrightarrow Y$ be a k ($k \geq 2$) times continuously Fréchet differentiable operator defined on a non-empty open convex domain Ω of a Banach space X*

with values in a Banach space Y and $f_0(t)$ the function defined in (3.12) with $w(t)$ replaced by $w_0(t)$. Suppose that conditions (V1)-(V2)-(V3) are satisfied. Then, the sequence $\{x_n\}$ given by Newton's method, converges to a solution x^* of $F(x) = 0$ starting at x_0. Moreover, $x_n, x^* \in \overline{B(x_0, t^* - t_0)}$ and

$$\|x^* - x_n\| \leq t^* - t_n, \quad \text{for all} \quad n = 0, 1, 2, \ldots,$$

where $\{t_n\}$ is defined in (4.1) with $f(t)$ replaced by $f_0(t)$.

It is known that, for example, there are operators that are Lipschitz continuous at some point of the domain Ω, but not at all the domain (see [39]). When this happens, we can guarantee the convergence of Newton's method using Theorem 3.34 rather than Theorem 3.29.

Theorem 3.34 also has an additional interest, since we can obtain better domains of existence and uniqueness of solution and error bounds, as we can see in the following example. Highlight that the improvements that we just comment relies on the idea, which is observed both in Kantorovich's polynomial (1.23) and Huang's polynomial (2.2), that the majorant function depends on the starting point. That is why we consider the general condition required for the k-th derivative of the operator involved in the starting point instead of the all domains Ω and, obviously $w_0(t) \leq w(t)$, so that $f_0(t) \leq f(t)$.

Remark 3.35. Taking into account conditions of this type, it is clear that the known relaxations of Huang's result are deduced obviously from Theorem 3.34. In particular, if F'' is center Lipschitz continuous at x_0, function (3.12) is reduced to the function

$$\zeta(t) = \frac{L_0}{6}t^3 + \frac{b_2}{2}t^2 - \frac{t}{\beta} + \frac{\eta}{\beta},$$

where L_0 is the Lipschitz constant, which is the same as that given in [37]; if F'' is center Hölder continuous at x_0, function (3.12) is reduced to the function

$$\varsigma(t) = \frac{H_0}{(2+p)(1+p)}t^{2+p} + \frac{b_2}{2}t^2 - \frac{t}{\beta} + \frac{\eta}{\beta},$$

where H_0 is the Hölder constant and $p \in [0,1]$, which is the same as that given in [37]; if F'' is center w_0-Lipschitz continuous at x_0, function (3.12) is reduced to the function

$$\vartheta(t) = \int_0^t \int_0^\theta w_0(\xi) \, d\xi \, d\theta + \frac{b_2}{2}t^2 - \frac{t}{\beta} + \frac{\eta}{\beta},$$

which is the same as that given in Section 2.1.3 (function (2.18)).

Example 3.36. Now, we apply Theorem 3.34 with $t_0 = 0$ to problem (3.14). For this, we have that $\|\mathfrak{F}^{(iv)}(x) - \mathfrak{F}^{(iv)}(x_0)\| \leq 360(2 + \|x_0\|)\|x - x_0\|$ with $x_0(s) = (2.9)s(1-s)$, $s \in [0,1]$, so that we can take $w_0(t) = 360(2 + \|x_0\|)t = 981t$ and

$$f_0(t) = (0.0049\ldots) - (0.3711\ldots)t + (2.1907\ldots)t^2 + (4.0939\ldots)t^3 + (45.2347\ldots)t^4,$$

where $f_0(t)$ is obtained from formula (3.12) with $w(t)$ replaced by $w_0(t)$ and $t_0 = 0$. Notice that $f_0(t)$ has two positive real zeros $z^* = 0.0145\ldots$ and $z^{**} = 0.1042\ldots$. Now, we observe that $w_0(t) \leq w(t)$ and $f_0(t) \leq f(t)$, see Figure 3.3. Therefore, the domains of existence and uniqueness of solutions of the boundary value problem (3.14) are better if $f_0(t)$ is considered

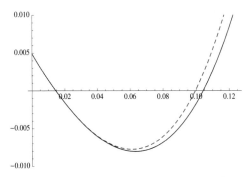

Figure 3.3: Functions $f(t)$, the dashed line, and $f_0(t)$, the solid line, of Example 3.36

n	$|z^* - z_n|$	$|t^* - t_n|$
0	$1.4528 \ldots \times 10^{-2}$	$1.4531 \ldots \times 10^{-2}$
1	$1.2852 \ldots \times 10^{-3}$	$1.2880 \ldots \times 10^{-3}$
2	$1.2818 \ldots \times 10^{-5}$	$1.2996 \ldots \times 10^{-5}$
3	$1.3097 \ldots \times 10^{-9}$	$1.3610 \ldots \times 10^{-9}$
4	$1.0408 \ldots \times 10^{-17}$	$1.5612 \ldots \times 10^{-17}$

Table 3.10: Error bounds obtained from the functions $f_0(t)$ and $f(t)$ of Example 3.36

instead of $f(t)$, since $z^* < t^*$ and $z^{**} > t^{**}$. Moreover, we also obtain better error bounds from the majorizing real sequence $\{z_n\}$, where $\{z_n\}$ is obtained when Newton's method is applied to the function $f_0(t)$, than from the majorizing real sequence $\{t_n\}$, where $\{t_n\}$ is obtained when Newton's method is applied to the function $f(t)$, as we can see in Table 3.10.

Finally, we use a discretization process to transform problem (3.14) into a finite-dimensional problem and look for an approximated solution of this problem. So, we transform problem (3.14) into a system of nonlinear equations by approximating the second derivative by a standard numerical formula. So, following the discretization process given in Section 1.2.2 for a boundary value problem of form (1.44)-(1.45), we reach system (1.52), given by

$$\mathbb{F}(\mathbf{x}) \equiv A\mathbf{x} - h^2 v_{\mathbf{x}} = 0, \tag{3.17}$$

where $\mathbb{F}$ is a function from $\mathbb{R}^m$ into $\mathbb{R}^m$ and

$$\mathbf{x} = \begin{pmatrix} x_1 \\ x_2 \\ \vdots \\ x_m \end{pmatrix}, \quad A = \begin{pmatrix} -2 & 1 & 0 & \cdots & 0 \\ 1 & -2 & 1 & \cdots & 0 \\ 0 & 1 & -2 & \cdots & 0 \\ \vdots & \vdots & \vdots & \ddots & \vdots \\ 0 & 0 & 0 & \cdots & -2 \end{pmatrix}, \quad v_{\mathbf{x}} = \begin{pmatrix} t_1 x_1^6 - 1 \\ t_2 x_2^6 - 1 \\ \vdots \\ t_m x_m^6 - 1 \end{pmatrix}.$$

Taking into account that we have previously considered the parabola $x_0(s) = (2.9)s(1 - s)$, $s \in [0, 1]$, as the starting function, we now choose this parabola discretized in $m + 1$ points in the interval $[0, 1]$ as the initial iterate for Newton's method. Then, if $m = 8$, after three iterations of Newton's method, the approximation given by the vector $\mathbf{x}^* =$

$(x_1^*, x_2^*, \ldots, x_8^*)^t$ shown in Table 3.11 is reached. In Table 3.12 we show the errors $\|\mathbf{x_n} - \mathbf{x}^*\|$ and the sequence $\{\|\mathbb{F}(\mathbf{x_n})\|\}$. From the last, we notice that the vector shown in Table 3.11 is a good approximation of the solution of system (3.17) with $m = 8$.

i	x_i^*	i	x_i^*
1	0.049382...	5	0.123456...
2	0.086419...	6	0.111111...
3	0.111111...	7	0.086419...
4	0.123456...	8	0.049382...

Table 3.11: Approximation of the solution $\mathbf{x}^*$ of system (3.17)

n	$\|\mathbf{x_n} - \mathbf{x}^*\|$	$\|\mathbb{F}(\mathbf{x_n})\|$
0	$5.9259 \ldots \times 10^{-1}$	$6.0183 \ldots \times 10^{-2}$
1	$1.8901 \ldots \times 10^{-2}$	$3.5197 \ldots \times 10^{-3}$
2	$6.1967 \ldots \times 10^{-8}$	$1.2675 \ldots \times 10^{-8}$

Table 3.12: Absolute errors and $\{\|\mathbb{F}(\mathbf{x_n})\|\}$

Chapter 4

Convergence conditions on the first derivative of the operator

In this chapter, we study the semilocal convergence of Newton's method under mild differentiability conditions on the operator F. If we observe the algorithm of Newton's method,

$$x_{n+1} = x_n - [F'(x_n)]^{-1}F(x_n), \quad n \geq 0, \quad \text{with } x_0 \text{ given,}$$

we see that the first derivative F' of the operator F is only involved. According to this, we can give only conditions on F', so that the semilocal convergence of the method is guaranteed. So, in the Newton-Kantorovich theorem the continuity condition on $F''(x)$ can be easily replaced by a Lipschitz continuity condition on $F'(x)$. This has been done by several authors, the first of which was, perhaps, Fenyö [32].

Note that the fact of requiring only conditions on F' leads to the fact that the method of majorizing sequences cannot be used to prove the semilocal convergence of Newton's method in all the situations that can be given. In Section 4.1, we see that if F' is Lipschitz continuous in the domain Ω, where the operator F is defined, and establish a semilocal convergence result for Newton's method as that given by Kantorovich in the Newton-Kantorovih theorem, where the second derivative F'' of the operator F is bounded in Ω. Once this is seen, the first relaxation that we think is that where the operator F' is Hölder continuous in Ω. And here is where the first problem appear with respect to Kantorovich's study, since we cannot use the method of majorizing sequences to prove the semilocal convergence of Newton's method, as we can see in Section 4.2. This leads us to prove the semilocal convergence of the method using a different technique from the method of majorizing sequences, which is based on recurrence relations [21], in the way of the original proof of Kantorovich of the Newton-Kantorovich theorem, Theorem 1.1. For a greater generality, we study the convergence of the method when F' is ω-Lipschitz continuous in Ω.

It is important to emphasise that we need to require some restriction on the domain Ω, so that the mild conditions mentioned above are satisfied in Ω, as we can see in Section 4.2. In addition, in Section 4.3, we delve more into this fact and center conditions on F' are required [39, 40, 42], so that the conditions are not satisfied at any two points x, y of the domain Ω, but they are satisfied at any point x of Ω and at the starting point x_0 of Newton's method.

Finally, in this chapter, we include some applications to nonlinear integral equations of Hammerstein-type, where results on existence and uniqueness of solutions are included from the theoretical significance of Newton's method.

© Springer International Publishing AG 2017

J.A. Ezquerro Fernández, M.Á. Hernández Verón, *Newton's Method: an Updated Approach of Kantorovich's Theory*, Frontiers in Mathematics, DOI 10.1007/978-3-319-55976-6_4

4.1   Operators with Lipschitz first-derivative

In this section, we prove the semilocal convergence of Newton's method under a Lipschitz condition on F'. For this, we use the method of majorizing sequences. Then, we obtain conditions for a scalar function, so that we can construct a majorant sequence for Newton's method in the Banach space X which leads us to a general result of semilocal convergence. In addition, we give a result on uniqueness of solution and study the R-order of convergence of Newton's method.

4.1.1   Existence of a majorizing sequence

As in Theorem 1.27, we need to construct a scalar function $f \in C^2([t_0, +\infty))$, with $t_0 \in \mathbb{R}$, such that $f''(t) \geq 0$ in $[t_0, +\infty)$ and satisfying the following conditions:

(N1) There exists $\Gamma_0 = [F'(x_0)]^{-1} \in \mathcal{L}(Y, X)$, for some $x_0 \in \Omega$, with $\|\Gamma_0\| \leq$
$$-\frac{1}{f'(t_0)} \text{ and } \|\Gamma_0 F(x_0)\| \leq -\frac{f(t_0)}{f'(t_0)}.$$

(N2) $\|F'(x) - F'(y)\| \leq f'(u) - f'(v)$, for $\|x - y\| \leq u - v$, $x, y \in \Omega$ and $u, v \in [t_0, +\infty)$.

Next, we prove the general semilocal convergence of Newton's method under conditions (N1)-(N2), by using the method of majorizing sequences, as Kantorovich does under conditions (a)-(b) of Theorem 1.27. For this, we construct a majorizing real sequence $\{t_n\}$,

$$t_0 \in I, \qquad t_n = N_f(t_{n-1}) = t_{n-1} - \frac{f(t_{n-1})}{f'(t_{n-1})}, \qquad n \in \mathbb{N}, \tag{4.1}$$

where $f : I \subseteq \mathbb{R} \longrightarrow \mathbb{R}$, that it is nondecreasing and convergent to a solution of the equation $f(t) = 0$, for Newton's sequence $\{x_n\}$ in the Banach space X.

To apply the method of majorizing sequences, the equation $f(t) = 0$ must have at least one solution $t^* > t_0$, so that scalar sequence (4.1) converges to t^* from t_0 under conditions (N1)-(N2). So, we first study the function f and give some properties of this function.

Theorem 4.1. Let $f \in C^2([t_0, +\infty))$, with $t_0 \in \mathbb{R}$, such that $f(t_0) > 0$, $f'(t_0) < 0$ and $f''(t) \geq 0$ in $[t_0, +\infty)$.

(a) If there exists a solution α of $f'(t) = 0$ such that $\alpha \in (t_0, +\infty)$, then α is the unique minimum of $f(t)$ in $[t_0, +\infty)$ and $f(t)$ is decreasing in $[t_0, \alpha)$.

(b) If $f(\alpha) \leq 0$, then $f(t) = 0$ has at least one solution in $[t_0, ++\infty)$. Moreover, if t^* is the smallest solution of $f(t) = 0$ in $[t_0, +\infty)$, then $t_0 < t^* \leq \alpha$.

Proof. First, as $f'(\alpha) = 0$ and $f''(t) \geq 0$ in $[t_0, +\infty)$, then α is a minimum of f in $[t_0, +\infty)$. Moreover, as f is convex in $[t_0, +\infty)$, then α is the unique minimum of f in $[t_0, +\infty)$ and f' is nondecreasing in $[t_0, +\infty)$.

Second, as $f'(t_0) < 0$, $f'(\alpha) = 0$ and f' is nondecreasing in $[t_0, +\infty)$, then $f'(t) < 0$ in $[t_0, \alpha)$, so that f is nonincreasing in $[t_0, \alpha)$.

Third, if $f(\alpha) < 0$, as $f(t_0) > 0$ and f is continuous, then f has at least one zero t^* in (t_0, α). As f is nonincreasing in $[t_0, \alpha)$, then t^* is the unique zero of f in (t_0, α).

Finally, if $f(\alpha) = 0$, then α is a double zero of f and $t^* = \alpha$.   ∎

After that, we establish the convergence of sequence (4.1) in the next result.

Theorem 4.2. *Let $\{t_n\}$ be the scalar sequence given in (4.1) with $f \in C^2([t_0, +\infty))$, $t_0 \in \mathbb{R}$ and such that $f''(t) \geq 0$ in $[t_0, +\infty)$. Suppose that conditions (N1)-(N2) hold and there exists a solution $\alpha \in (t_0, +\infty)$ of $f'(t) = 0$ such that $f(\alpha) \leq 0$. Then, (4.1) is a nondecreasing sequence that converges to t^*.*

Observe that if there exists such function f, then it has the same properties as Kantorovich's polynomial (1.23), since $f(t) > 0$, $f'(t) < 0$ and $f''(t) > 0$ in (t_0, α), so that the convergence of the real sequence given in (4.1) is then guaranteed in a way analogous to that of sequence (1.22) and such as we have seen in Theorem 1.21.

Now, we see that (4.1) is a majorizing sequence of Newton's sequence $\{x_n\}$ defined in the Banach space X, as a consequence of the following result.

Lemma 4.3. *Suppose that there exists $f \in C^2([t_0, +\infty))$, with $t_0 \in \mathbb{R}$, such that $f''(t) \geq 0$ in $[t_0, +\infty)$ and conditions (N1)-(N2) are satisfied. Suppose also that $x_n \in \Omega$, for all $n \geq 0$, and $f(\alpha) \leq 0$, where α is a solution of $f'(t) = 0$ in $(t_0, +\infty)$. Then, for all $n \geq 1$, we have the following:*

(i_n) *There exists $\Gamma_n = [F'(x_n)]^{-1}$ and $\|\Gamma_n\| \leq -\dfrac{1}{f'(t_n)}$,*

(ii_n) *$\|F(x_n)\| \leq f(t_n)$,*

(iii_n) *$\|x_{n+1} - x_n\| \leq t_{n+1} - t_n$.*

Proof. From (N1), we observe

$$\|x_1 - x_0\| = \|\Gamma_0 F(x_0)\| \leq -\frac{f(t_0)}{f'(t_0)} = t_1 - t_0 < t^* - t_0.$$

After that, we prove by mathematical induction on n that items (i_n)-(ii_n)-(iii_n) are true. The first step, $n = 1$, is analogous to which we do for the inductive step, no more to consider conditions (N1)-(N2). For the inductive step, we suppose that (i_n)-(ii_n)-(iii_n) are true for $n = 1, 2, \ldots, j - 1$ and prove items (i_j)-(ii_j)-(iii_j).

First, from

$$\|I - \Gamma_{j-1} F'(x_j)\| \leq \|\Gamma_{j-1}\| \|F'(x_{j-1}) - F'(x_j)\|$$
$$\leq -\frac{1}{f'(t_{j-1})} (f'(t_j) - f'(t_{j-1}))$$
$$= 1 - \frac{f'(t_j)}{f'(t_{j-1})}$$

and $\frac{f'(t_j)}{f'(t_{j-1})} \in (0, 1)$, since $t_{j-1} < t_j \leq t^*$ and $f''(t) > 0$ in $[t_0, +\infty)$, we have $\|I - \Gamma_{j-1} F'(x_j)\| < 1$. As a consequence, by the Banach lemma on invertible operators, there exists Γ_j and $\|\Gamma_j\| \leq -\frac{1}{f'(t_j)}$.

Second, taking into account that $F(x_{j-1}) + F'(x_{j-1})(x_j - x_{j-1}) = 0$ and Taylor's series, it follows

$$\|F(x_j)\| = \left\| \int_0^1 \left(F'(x_{j-1} + \tau(x_j - x_{j-1})) - F'(x_{j-1}) \right) (x_j - x_{j-1}) \, d\tau \right\|$$

$$\leq \int_0^1 \|F'(x_{j-1} + \tau(x_j - x_{j-1})) - F'(x_{j-1})\| \, \|x_j - x_{j-1}\| \, d\tau$$

$$\leq \int_0^1 \left(f'(t_{j-1} + \tau(t_j - x_{j-1})) - f'(x_{j-1}) \right) (t_j - t_{j-1}) \, d\tau$$

$$= f(t_j).$$

Third, $\|x_{j+1} - x_j\| \leq \|\Gamma_j\| \|F(x_j)\| \leq -\dfrac{f(t_j)}{f'(t_j)} = t_{j+1} - t_j$.

Finally, as a consequence, items (i_n)-(ii_n)-(iii_n) are true for all positive integers n by mathematical induction. ∎

Then, we are ready to prove the general semilocal convergence of Newton's method in the Banach space X under conditions (N1)-(N2).

Theorem 4.4. (General semilocal convergence) *Let $F : \Omega \subseteq X \longrightarrow Y$ be a continuously Fréchet differentiable operator defined on a non-empty open convex domain Ω of a Banach space X with values in a Banach space Y. Suppose that there exists $f \in C^2([t_0, +\infty))$, with $t_0 \in \mathbb{R}$, such that $f''(t) \geq 0$ in $[t_0, +\infty)$ and (N1)-(N2) are satisfied. If there exists $\alpha \in (t_0, +\infty)$ such that $f'(\alpha) = 0$ and $f(\alpha) \leq 0$, and $B(x_0, t^* - t_0) \subset \Omega$, then Newton's sequence $\{x_n\}$ converges to a solution x^* of $F(x) = 0$ starting at x_0. Moreover, $x_n, x^* \in \overline{B}(x_0, t^* - t_0)$ and*

$$\|x^* - x_n\| \leq t^* - t_n, \quad \text{for all} \quad n = 0, 1, 2, \ldots,$$

where $\{t_n\}$ is defined in (4.1).

Proof. From (N1)-(N2), it is clear that x_1 is well-defined and $\|x_1 - x_0\| = t_1 - t_0 < t^* - t_0$, so that $x_1 \in B(x_0, t^* - t_0) \subset \Omega$.

We now suppose that x_j is well-defined and $x_j \in B(x_0, t^* - t_0)$ for $j = 1, 2, \ldots, n - 1$.

After that, by the last lemma, the operator Γ_{n-1} exists and $\|\Gamma_{n-1}\| \leq -\dfrac{1}{f'(t_{n-1})}$. In addition, x_n is well-defined. Moreover, since $\|x_{j+1} - x_j\| \leq t_{j+1} - t_j$ for $j = 1, 2, \ldots, n - 1$, we have

$$\|x_n - x_0\| \leq \sum_{j=0}^{n-1} \|x_{j+1} - x_j\| \leq \sum_{j=0}^{n-1} (t_{j+1} - t_j) = t_n - t_0 < t^* - t_0,$$

so that $x_n \in B(x_0, t^* - t_0)$. Therefore, Newton's sequence $\{x_n\}$ is well-defined and $x_n \in B(x_0, t^* - t_0)$ for all $n \geq 0$.

By the last lemma, we also have that $\|F(x_n)\| \leq f(t_n)$ and

$$\|x_{n+1} - x_n\| \leq \|\Gamma_n\| \|F(x_n)\| \leq -\dfrac{f(t_n)}{f'(t_n)} = t_{n+1} - t_n,$$

for all $n \geq 0$, so that $\{t_n\}$ is a majorizing real sequence of $\{x_n\}$. Moreover, as $\{t_n\}$ is convergent, then $\lim_n t_n = t^*$, so that, if $x^* = \lim_n x_n$, then $\|x^* - x_n\| \leq t^* - t_n$, for all $n = 0, 1, 2, \ldots$ Furthermore, as $\|F(x_n)\| \leq f(t_n)$, for all $n = 0, 1, 2, \ldots$, then, taking into account $n \to +\infty$, it follows $F(x^*) = 0$ by the continuity of F. ∎

After proving the existence of a majorizing sequence of Newton's method in X and locating the solution x^*, we prove the uniqueness of x^*.

Theorem 4.5. *Under the hypotheses of Theorem 4.4, if the scalar function $f(t)$ has two real zeros t^* and t^{**} such that $t_0 < t^* \leq t^{**}$, then the solution x^* is unique in $B(x_0, t^{**} - t_0) \cap \Omega$ if $t^* < t^{**}$ or in $\overline{B(x_0, t^* - t_0)}$ if $t^{**} = t^*$.*

Proof. Suppose that $t^* < t^{**}$ and y^* is another solution of $F(x) = 0$ in $B(x_0, t^{**} - t_0) \cap \Omega$. From

$$0 = F(y^*) - F(x^*) = \int_{x^*}^{y^*} F'(x)dx = \int_0^1 F'(x^* + \tau(y^* - x^*)) \, d\tau (y^* - x^*),$$

it follows that $x^* = y^*$, provided that the operator $\int_0^1 F'(x^* + \tau(y^* - x^*)) \, d\tau$ is invertible. To prove this, we prove equivalently that there exists the operator J^{-1}, where $J = \Gamma_0 \int_0^1 F'(x^* + \tau(y^* - x^*)) \, d\tau$. Indeed, as

$$\|I - J\| \leq \|\Gamma_0\| \int_0^1 \|F'(x^* + \tau(y^* - x^*)) - F'(x_0)\| \, d\tau$$

$$< -\frac{1}{f'(t_0)} \int_0^1 (f'(t^* + \tau(t^{**} - t^*)) - f'(t_0)) \, d\tau$$

$$= 1,$$

since $\|x^* + \tau(y^* - x^*) - x_0\| \leq t^* + \tau(t^{**} - t^*) - t_0$ with $\tau \in [0, 1]$, the operator J^{-1} exists by the Banach lemma on invertible operators.

If $t^{**} = t^*$ and y^* is another solution of $F(x) = 0$ in $\overline{B(x_0, t^* - t_0)}$, then $\|y^* - x_0\| \leq t^* - t_0$. We now suppose that $\|y^* - x_j\| \leq t^* - t_j$ for $j = 0, 1, \ldots, n$. In addition,

$$\|y^* - x_{n+1}\| = \|\Gamma_n (F(y^*) - F(x_n) - F'(x_n)(y^* - x_n))\|$$

$$= \left\| \Gamma_n \int_{x_n}^{y^*} (F'(x) - F'(x_n)) \, dx \right\|$$

$$= \left\| \Gamma_n \int_0^1 (F'(x_n + \tau(y^* - x_n)) - F'(x_n)) \, d\tau \, (y^* - x_n) \right\|$$

$$\leq \|\Gamma_n\| \int_0^1 \|F'(x_n + \tau(y^* - x_n)) - F'(x_n)\| \, d\tau \, \|y^* - x_n\|$$

$$\leq -\frac{1}{f'(t_n)} \int_0^1 (f'(t_n + \tau(t^* - t_n)) - f'(t_n)) \, d\tau \, \|y^* - x_n\|.$$

Besides, as

$$t^* - t_{n+1} = t^* - t_n + \frac{f(t_n)}{f'(t_n)}$$

$$= -\frac{1}{f'(t_n)} (f(t^*) - f(t_n) + f'(t_n)(t^* - t_n))$$

$$= -\frac{1}{f'(t_n)} \int_0^1 (f'(t_n + \tau(t^* - t_n)) - f'(t_n))(t^* - t_n) \, d\tau,$$

we have

$$\|y^* - x_{n+1}\| \leq \frac{t^* - t_{n+1}}{t^* - t_n} \|y^* - x_n\| \leq t^* - t_{n+1}.$$

Now, since $\lim_n t_n = t^*$, the uniqueness of x^* follows immediately. ∎

4.1.2   Existence of a majorant function

Now, we suppose that conditions (N1)-(N2) are reduced respectively to the following conditions:

> (P1) There exists $\Gamma_0 = [F'(x_0)]^{-1} \in \mathcal{L}(Y, X)$, for some $x_0 \in \Omega$, with $\|\Gamma_0\| \leq \beta$ and $\|\Gamma_0 F(x_0)\| \leq \eta$.
>
> (P2) There exists a constant $M \geq 0$ such that $\|F'(x) - F'(y)\| \leq M\|x - y\|$, for $x, y \in \Omega$.

Observe that condition (A2) of Kantorovich implicates (P2). On the other hand, to find a real function from conditions (P1)-(P2), we have

$$\|F'(x) - F'(y)\| \leq M\|x - y\| \leq M(u - v) = f'(u) - f'(v)$$

if $\|x - y\| \leq u - v$, so that

$$\int_u^v M\, d\tau = \int_u^v f''(\tau)\, d\tau.$$

Therefore, it is enough to solve the following initial value problem:

$$\begin{cases} y''(t) - M = 0, \\ y(t_0) = \dfrac{\eta}{\beta}, \quad y'(t_0) = -\dfrac{1}{\beta}, \end{cases}$$

whose solution is the quadratic polynomial

$$f(t) = \frac{M}{2}(t - t_0)^2 - \frac{t - t_0}{\beta} + \frac{\eta}{\beta}, \qquad (4.2)$$

which is reduced to Kantorovich's polynomial (1.23) if $t_0 = 0$.

4.1.3   Semilocal convergence

Notice that polynomial (4.2) satisfies the hypotheses of Theorem 4.4 and, as a consequence, the semilocal convergence of Newton's method is guaranteed in X. In particular, from Theorem 4.4, we obtain the following variant of the Newton-Kantorovich theorem, which is the best known, was given by Ortega in [63] and is also called the Newton-Kantorovich theorem.

Theorem 4.6. *Let $F : \Omega \subseteq X \longrightarrow Y$ be a continuously Fréchet differentiable operator defined on a non-empty open convex domain Ω of a Banach space X with values in a Banach space Y. Suppose that conditions (P1)-(P2) are satisfied. If $M\beta\eta \leq \frac{1}{2}$ and $B(x_0, s^*) \subset \Omega$, where $s^* = \frac{1 - \sqrt{1 - 2M\beta\eta}}{M\beta}$, then Newton's sequence $\{x_n\}$ converges to a solution x^* of the equation $F(x) = 0$, starting at x_0, and $x_n, x^* \in \overline{B(x_0, s^*)}$, for all $n \geq 0$. Moreover, if $M\beta\eta < \frac{1}{2}$, x^* is the unique solution of $F(x) = 0$ in $B(x_0, s^{**}) \cap \Omega$, where $s^{**} = \frac{1 + \sqrt{1 - 2M\beta\eta}}{M\beta}$, and if $M\beta\eta = \frac{1}{2}$, x^* is unique in $\overline{B(x_0, s^*)}$. Furthermore,*

$$\|x_{n+1} - x_n\| \leq s_{n+1} - s_n \quad and \quad \|x^* - x_n\| \leq s^* - s_n, \quad for\ all\ \ n \geq 0,$$

where $s_n = s_{n-1} - \frac{p(s_{n-1})}{p'(s_{n-1})}$, $n \in \mathbb{N}$, $s_0 = 0$ and $p(s)$ is Kantorovich's polynomial (1.23).

In addition, we can also see the quadratic convergence of Newton's method under conditions (P1)-(P2). First, we observe that

$$p(s) = \frac{M}{2}(s^* - s)(s^{**} - s). \tag{4.3}$$

Sedond, we give the following result which provides some error estimates that lead to the quadratic convergence of Newton's method. The proof of this result follows from Ostrsowski's technique [65] and is completely analogous to that of Theorem 2.18.

Theorem 4.7. *Under the hypotheses of Theorem 4.6, we have the following:*

(a) *If $s^* < s^{**}$, then $\theta = \frac{s^*}{s^{**}} < 1$ and*

$$s^* - s_n = \frac{(s^{**} - s^*)\theta^{2^n}}{1 - \theta^{2^n}}, \quad n \geq 0.$$

(b) *If $s^* = s^{**}$, then*

$$s^* - s_n = \frac{s^*}{2^n}, \quad n \geq 0.$$

Proof. From (4.3), it follows that $p(s_n) = a_n b_n$ and $p'(s_n) = -\frac{M}{2}(a_n + b_n)$ with $a_n = s^* - s_n$ and $b_n = s^{**} - s_n$, for all $n \geq 0$. Moreover,

$$a_{n+1} = s^* - s_{n+1} = s^* - s_n + \frac{p(s_n)}{p'(s_n)} = \frac{a_n{}^2}{(a_n + b_n)}, \quad b_{n+1} = \frac{b_n{}^2}{(a_n + b_n)}.$$

As a consequence,

$$\frac{a_{n+1}}{b_{n+1}} = \left(\frac{a_n}{b_n}\right)^2 = \cdots = \left(\frac{a_0}{b_0}\right)^{2^{n+1}} = \left(\frac{s^*}{s^{**}}\right)^{2^{n+1}} = \theta^{2^{n+1}}.$$

If $s^* < s^{**}$, then $b_{n+1} = (s^{**} - s^*) + a_{n+1}$ and $a_{n+1} = b_{n+1}\theta^{2^{n+1}}$. Therefore,

$$s^* - s_{n+1} = a_{n+1} = \frac{(s^{**} - s^*)\theta^{2^{n+1}}}{1 - \theta^{2^{n+1}}}.$$

If $s^* = s^{**} = \frac{1}{M\beta}$, then $a_n = b_n$ and $a_{n+1} = \frac{a_n}{2}$. Therefore,

$$a_{n+1} = \frac{s^*}{2^{n+1}}.$$

The proof is complete. ∎

From the last theorem, it follows that the convergence of Newton's method, under conditions (P1)-(P2), is quadratic if $s^* < s^{**}$ and linear if $s^* = s^{**}$.

4.2   Operators with ω-Lipschitz first-derivative

When the conditions required to the operator involved F are relaxed to prove the semilocal convergence of Newton's method, the natural next step is to consider operators with Hölder continuous first Fréchet derivative:

$$\|F'(x) - F'(y)\| \le K\|x - y\|^p, \quad x, y \in \Omega, \quad p \in [0, 1].$$

Vertgeim [79] was the first to weaken the continuity condition on $F''(x)$ to the last Hölder condition.

Now, if we try to construct a majorant sequence from a real function f, as we have done in Section 4.1 for operators with Lipschitz first-derivative, we obtain

$$\|F'(x) - F'(y)\| \le K\|x - y\|^p \le K(u - v)^p$$

if $\|x - y\| \le u - v$ and look for a real function f such that $K(u - v)^p = f'(u) - f'(v)$. If we first do the same as Kantorovich and look for a polynomial, we see that this is not possible. Second, if we use the technique of the initial value problem, then

$$\int_v^u f''(t)\, dt = K \int_0^{u-v} p\, t^{p-1}\, dt,$$

which leads to a differential problem that is difficult to solve. These two problems lead us to introduce a new technique for proving the semilocal convergence of Newton's method which is different from the method of majorizing sequences and is based on proving a certain system of recurrence relations. Kantorovich himself, before using the majorant principle to prove the semilocal convergence of Newton's method, already used a technique based on recurrence relations to prove the semilocal convergence of the method (see Theorem 1.1).

As we have already indicated above, a generalization of condition (P2), under which the semilocal convergence of Newton's method is also usually studied, is when the first derivative F' is Hölder continuous in Ω. That is (as we can see in [43, 56, 75]):

(P3)   There exist two constants $K \ge 0$ and $p \in [0, 1]$ such that $\|F'(x) - F'(y)\| \le K\|x - y\|^p$, for $x, y \in \Omega$.

Observe that if $p = 1$ in (P3), F' is Lipschitz continuous in Ω. In addition, we can also find F' defined as a combination of operators such that F' is Lipschitz or Hölder continuous in Ω (see Section 2.1.3). That is:

(P4)   There exist $2m$ constants $K_1, K_2, \ldots, K_m \ge 0$ and $p_1, p_2, \ldots, p_m \in [0, 1]$ such that $\|F'(x) - F'(y)\| \le \sum_{i=1}^m K_i\|x - y\|^{p_i}$, for $x, y \in \Omega$.

To give sufficient generality to all the above conditions on the first derivative F', we consider the following condition:

(P5)   There exist two continuous and nondecreasing functions $\omega : [0, +\infty) \longrightarrow \mathbb{R}$ and $h : [0, 1] \longrightarrow \mathbb{R}$ such that $\omega(0) = 0$, $\omega(tz) \le h(t)\omega(z)$, with $t \in [0, 1]$, $z \in [0, +\infty)$ and $\|F'(x) - F'(y)\| \le \omega(\|x - y\|)$, for $x, y \in \Omega$.

We then say that F' is ω-Lipschitz continuous in Ω. Obviously, we obtain (P2) if $\omega(z) = Mz$ and $h(t) = t$, (P3) if $\omega(z) = Kz^p$ and $h(t) = t^p$, and (P4) if $\omega(z) = \sum_{i=1}^m K_i z^{p_i}$ and $h(t) = t^p$ where $p = \min_{i=1,2,\ldots,m}\{p_i\}$ and $p_i \in [0, 1]$, for all $i = 1, 2, \ldots, m$. Moreover, note that condition $\omega(tz) \le h(t)\omega(z)$, for $t \in [0, 1]$, does not involve any restriction, since h always exists, such that $h(t) = 1$, for $t \in [0, 1]$, as a consequence of ω is a nondecreasing function.

4.2.1 Semilocal convergence using recurrence relations

In the following, we analyse the semilocal convergence of Newton's method, under condition (P5), by using the technique based on recurrence relations. From some real parameters, a system of three recurrence relations is now constructed in which three sequences of positive real numbers are involved. The convergence of Newton's sequence in the Banach space X is then guaranteed from it.

4.2.1.1 Recurrence relations

We suppose:

(Q1) There exists $\Gamma_0 = [F'(x_0)]^{-1} \in \mathcal{L}(Y, X)$, for some $x_0 \in \Omega$, with $\|\Gamma_0\| \leq \beta$ and $\|\Gamma_0 F(x_0)\| \leq \eta$.

(Q2) There exist two continuous and nondecreasing functions $\omega : [0, +\infty) \longrightarrow \mathbb{R}$ and $h : [0, 1] \longrightarrow \mathbb{R}$ such that $\omega(0) = 0$, $\omega(tz) \leq h(t)\omega(z)$, with $t \in [0, 1]$, $z \in [0, +\infty)$ and $\|F'(x) - F'(y)\| \leq \omega(\|x - y\|)$, for $x, y \in \Omega$.

Note that inequality $\omega(tz) \leq h(t)\omega(z)$, with $t \in [0, 1]$ and $z \in [0, +\infty)$, does not involve any restriction, since h always exists, such that $h(t) = 1$, as a consequence of ω is a nondecreasing function. We use it to sharpen the bounds that we obtain for particular expressions, as we will see later.

We now denote $a_0 = \eta$, $b_0 = \beta\omega(a_0)$ and define the following scalar sequences:

$$t_n = I_h b_n f(b_n), \quad n \geq 0, \tag{4.4}$$
$$a_n = t_{n-1} a_{n-1}, \quad n \geq 1, \tag{4.5}$$
$$b_n = h(t_{n-1}) b_{n-1} f(b_{n-1}), \quad n \geq 1, \tag{4.6}$$

where $I_h = \int_0^1 h(t)\, dt$ and

$$f(t) = \frac{1}{1-t}. \tag{4.7}$$

Observe that we consider the case $b_0 > 0$, since if $b_0 = 0$, a trivial problem results.

Next, we see the following recurrence relations that sequences $\{x_n\}$, (4.5) and (4.6) satisfy:

$$\|\Gamma_1\| = \|[F'(x_1)]^{-1}\| \leq f(b_0)\|\Gamma_0\|, \tag{4.8}$$
$$\|x_2 - x_1\| \leq a_1, \tag{4.9}$$
$$\|\Gamma_1\|\omega(a_1) \leq b_1. \tag{4.10}$$

For this, we assume that
$$x_1 \in \Omega \quad \text{and} \quad b_0 < 1.$$

As Γ_0 exists, by the Banach lemma on invertible operators, we have that there exists Γ_1 and

$$\|\Gamma_1\| \leq \frac{\|\Gamma_0\|}{1 - \|I - \Gamma_0 F'(x_1)\|} \leq f(b_0)\|\Gamma_0\|,$$

since

$$\|I - \Gamma_0 F'(x_1)\| \leq \|\Gamma_0\|\|F'(x_0) - F'(x_1)\| \leq \beta\omega(a_0) = b_0 < 1.$$

From Taylor's series and $\{x_n\}$, it follows

$$\|F(x_1)\| = \left\| \int_0^1 (F'(x_0 + \tau(x_1 - x_0)) - F'(x_0))(x_1 - x_0) \, d\tau \right\|$$
$$\leq \left(\int_0^1 \omega(ta_0) \, dt \right) \|x_1 - x_0\|$$
$$\leq I_h \omega(a_0) \|x_1 - x_0\|.$$

Thus,

$$\|x_2 - x_1\| \leq \|\Gamma_1\| \|F(x_1)\| \leq I_h b_0 f(b_0) \|x_1 - x_0\| \leq t_0 a_0 = a_1,$$
$$\|\Gamma_1\| \omega(a_1) \leq f(b_0) \|\Gamma_0\| \omega(t_0 a_0) \leq f(b_0) \beta h(t_0) \omega(a_0) = h(t_0) b_0 f(b_0) = b_1.$$

Later, in Theorem 4.9 we generalize (4.8)-(4.9)-(4.10) to every point of the sequence $\{x_n\}$ and prove that $\{x_n\}$ is a Cauchy sequence. For this, we first investigate the scalar sequences $\{t_n\}$, $\{a_n\}$ and $\{b_n\}$ at the beginning of the following section.

4.2.1.2 Analysis of the scalar sequences

To analyse the real sequences (4.4)-(4.5)-(4.6), so that the convergence of the sequence $\{x_n\}$ is guaranteed in X. It suffices to see that $x_{n+1} \in \Omega$ and $b_n < 1$, for all $n \geq 1$. Next, we generalize the previous recurrence relations, so that we can prove that $\{x_n\}$ is a Cauchy sequence. First, we give a technical lemma.

Lemma 4.8. *Let f be the scalar function defined in (4.7). If*

$$b_0 = \beta \omega(a_0) \leq \frac{1}{1 + I_h} \qquad and \qquad b_0 < 1 - h(t_0), \tag{4.11}$$

with $I_h = \int_0^1 h(t) \, dt$, then

(a) *the sequences $\{t_n\}$, $\{a_n\}$ and $\{b_n\}$ are strictly decreasing,*

(b) *$t_n < 1$ and $b_n < 1$, for all $n \geq 0$.*

If $b_0 = 1 - h(t_0) < \frac{1}{1+I_h}$, then $t_n = t_0 < 1$ and $b_n = b_0 < 1$, for all $n \geq 1$.

Proof. Item (a) is proved by mathematical induction on n. As $b_0 < 1 - h(t_0)$, then $b_1 < b_0$ and $t_1 < t_0$, since f is increasing. Moreover, $a_1 < a_0$ as a consequence of $t_0 < 1$ and b_0 satisfies (4.11). Next, we suppose that $b_i < b_{i-1}$, $t_i < t_{i-1}$ and $a_i < a_{i-1}$, for all $i = 1, 2, \ldots, n$. Thus,

$$b_{n+1} = h(t_n) b_n f(b_n) < h(t_0) b_n f(b_0) < b_n,$$
$$t_{n+1} = I_h b_n f(b_n) < I_h b_{n-1} f(b_{n-1}) = t_n,$$
$$a_{n+1} = t_n a_n < t_0 \, a_n < a_n,$$

since f and h are increasing in $[0, 1)$. Consequently, the sequences $\{t_n\}$, $\{a_n\}$ and $\{b_n\}$ are strictly decreasing.

To see item (b), we have $t_n < t_0 < 1$ and $b_n < b_0 < 1$, for all $n \geq 0$, by item (a) and (4.11).

Finally, if $b_0 = 1 - h(t_0)$, it follows that $h(t_0) f(b_0) = 1$, and therefore, $b_n = b_0 = 1 - h(t_0) < 1$, for all $n \geq 0$. Moreover, if $b_0 < \frac{1}{1+I_h}$, then we have $t_n = t_0 < 1$, for all $n \geq 0$. ∎

4.2.1.3 Semilocal convergence

We are now ready to prove a semilocal convergence result for Newton's method when it is applied to operators that satisfy conditions (Q1)-(Q2).

Theorem 4.9. *Let $F : \Omega \subseteq X \longrightarrow Y$ be a continuously Fréchet differentiable operator defined on a non-empty open convex domain Ω of a Banach space X with values in a Banach space Y. Suppose that (Q1)-(Q2) are satisfied. Suppose also that (4.11) is satisfied and $B(x_0, R) \subset \Omega$, where $R = \frac{a_0}{1-t_0}$. Then, Newton's sequence $\{x_n\}$ converges to a solution x^* of $F(x) = 0$ starting at x_0 and $x_n, x^* \in \overline{B(x_0, R)}$.*

Proof. First, we prove the following four items for the sequence $\{x_n\}$ and $n \geq 1$:

(i_n) There exists $\Gamma_n = [F'(x_n)]^{-1}$ and $\|\Gamma_n\| \leq f(b_{n-1})\|\Gamma_{n-1}\|$,

(ii_n) $\|x_{n+1} - x_n\| \leq a_n$,

(iii_n) $\|\Gamma_n\|\omega(a_n) \leq b_n$,

(iv_n) $x_{n+1} \in \Omega$.

Notice that $x_1 \in \Omega$, since $\eta < R$. Then, from (4.8)-(4.9)-(4.10) and

$$\|x_2 - x_0\| \leq \|x_2 - x_1\| + \|x_1 - x_0\| \leq a_1 + a_0,$$

it follows that items (i_n)-(ii_n)-(iii_n)-(iv_n) are satisfied for $n = 1$. If we now suppose that items (i_n)-(ii_n)-(iii_n)-(iv_n) are true for some $n = 1, 2, \ldots, j$, we see that items (i_{j+1})-(ii_{j+1})-(iii_{j+1})-(iv_{j+1}) are also satisfied. We take into account that $t_j < 1$ and $b_j < 1$, for all $j \geq 0$.
(i_{j+1}): Observe

$$\begin{aligned}
\|I - \Gamma_j F'(x_{j+1})\| &\leq \|\Gamma_j\|\omega(\|x_{j+1} - x_j\|) \\
&\leq f(b_{j-1})\|\Gamma_{j-1}\|\omega(a_j) \\
&= f(b_{j-1})\|\Gamma_{j-1}\|\omega(t_{j-1}a_{j-1}) \\
&\leq f(b_{j-1})\|\Gamma_{j-1}\|h(t_{j-1})\omega(a_{j-1}) \\
&\leq h(t_{j-1})b_{j-1}f(b_{j-1}) \\
&= b_j \\
&< 1,
\end{aligned}$$

since $\{b_n\}$ is decreasing, $b_0 < \frac{1}{1+I_h}$ and $t_{j-1} < 1$. Then, by the Banach lemma on invertible operators, there exists Γ_{j+1} and

$$\|\Gamma_{j+1}\| \leq \frac{\|\Gamma_j\|}{1 - b_j} = f(b_j)\|\Gamma_j\|.$$

(ii_{j+1}): By Taylor's series and $\{x_n\}$ it follows, as for (4.9), that

$$\begin{aligned}
\|F(x_{j+1})\| &= \left\| \int_0^1 (F'(x_j + \tau(x_{j+1} - x_j)) - F'(x_j))(x_{j+1} - x_j) \, d\tau \right\| \\
&\leq \left(\int_0^1 \omega(\tau\|x_{j+1} - x_j\|) \, d\tau \right) \|x_{j+1} - x_j\| \\
&\leq I_h \omega(a_j) \, a_j.
\end{aligned}$$

Therefore,

$$\|x_{j+2} - x_{j+1}\| \leq f(b_j)\|\Gamma_j\|I_h\omega(a_j)a_j \leq I_h b_j f(b_j)a_j = t_j a_j = a_{j+1}.$$

(iii_{j+1}): The inequality

$$\|\Gamma_{j+1}\|\omega(a_{j+1}) \leq b_{j+1}$$

follows immediately.

(iv_{j+1}): In addition,

$$
\begin{aligned}
\|x_{j+2} - x_0\| \quad &\leq \quad \|x_{j+2} - x_{j+1}\| + \|x_{j+1} - x_0\| \\
&\leq \quad a_{j+1} + \sum_{k=0}^{j} a_k \\
&= \quad a_0\left(1 + \sum_{k=0}^{j}\left(\prod_{p=0}^{k} t_k\right)\right) \\
(\{t_k\}\searrow) \quad &< \quad a_0\left(1 + \sum_{k=0}^{j} t_0^{k+1}\right) \\
&= \quad \frac{1 - t_0^{j+2}}{1 - t_0}a_0 \\
&< \quad \frac{a_0}{1 - t_0} \\
&= \quad R.
\end{aligned}
$$

As a consequence, $x_{j+2} \in B(x_0, R) \subset \Omega$. The induction is complete.

Second, we prove that $\{x_n\}$ is a Cauchy sequence. For this, we have, for $m \geq 1$ and $n \geq 1$,

$$\|x_{n+m} - x_n\| \leq \sum_{j=n}^{n+m-1}\|x_{j+1} - x_j\| \overset{(ii_j)}{\leq} \sum_{j=n}^{n+m-1} a_j \leq \frac{1 - t_0^m}{1 - t_0}a_0 t_0^n.$$

Thus, $\{x_n\}$ is a Cauchy sequence.

Third, we show that x^* is a solution of the equation $F(x) = 0$. As $\|\Gamma_n F(x_n)\| \to 0$ when $n \to +\infty$, if we take into account that

$$\|F(x_n)\| \leq \|F'(x_n)\|\|\Gamma_n F(x_n)\| = \|F'(x_n)\|\|x_{n+1} - x_n\|$$

and $\{\|F'(x_n)\|\}$ is bounded, we obtain $\|F(x_n)\| \to 0$ when $n \to +\infty$, since

$$\|F'(x_n)\| \overset{(Q2)}{\leq} \|F'(x_0)\| + \omega(\|x_n - x_0\|) < \|F'(x_0)\| + \omega(R).$$

Therefore, by the continuity of F in $\overline{B(x_0, R)}$, it follows $F(x^*) = 0$.   ∎

Remark 4.10. If $b_0 = 1 - h(t_0) < \frac{1}{1+I_h}$, it follows, similarly to the previous theorem, that Newton's sequence is convergent in the Banach space X.

4.2.2 Uniqueness of solution and order of convergence

Once we have proved the semilocal convergence of Newton's method and located the solution x^*, we prove the uniqueness of x^*.

Theorem 4.11. *Let $F : \Omega \subseteq X \longrightarrow Y$ be a continuously Fréchet differentiable operator defined on a non-empty open convex domain Ω of a Banach space X with values in a Banach space Y. If conditions(Q1)-(Q2) are satisfied, then the solution x^* is unique in $B(x_0, r) \cap \Omega$, where r is the bigest positive real solution of the equation*

$$\frac{\beta}{r - R} (\Upsilon(r) - \Upsilon(R)) = 1, \tag{4.12}$$

where $\Upsilon(r) = \int_0^r \omega(\tau) \, d\tau$.

Proof. Suppose that y^* is another solution of $F(x) = 0$ in $B(x_0, r) \cap \Omega$. Then, from the approximation

$$0 = F(y^*) - F(x^*) = \int_{x^*}^{y^*} F'(x) dx = \int_0^1 F'(x^* + \tau(y^* - x^*)) \, d\tau (y^* - x^*),$$

it follows $x^* = y^*$, provided that the operator $\int_0^1 F'(x^* + \tau(y^* - x^*)) \, d\tau$ is invertible. For this, we prove equivalently that there exists the operator J^{-1}, where $J = \Gamma_0 \int_0^1 F'(x^* + \tau(y^* - x^*)) \, d\tau$. Indeed, as

$$\|I - J\| \leq \|\Gamma_0\| \int_0^1 \|F'(x^* + \tau(y^* - x^*)) - F'(x_0)\| \, d\tau$$
$$\leq \beta \int_0^1 \omega(\|x_0 - x^* - \tau(y^* - x^*)\|) \, d\tau$$
$$\leq \beta \int_0^1 \omega(\|(1 - \tau)(x_0 - x^*) - \tau(y^* - x_0)\|) \, d\tau$$
$$\leq \beta \int_0^1 \omega((1 - \tau)\|x^* - x_0\| + \tau\|y^* - x_0\|) \, d\tau \tag{4.13}$$
$$< \beta \int_0^1 \omega(R + \tau(r - R)) \, d\tau$$
$$= \frac{\beta}{r - R} \int_R^r \omega(\tau) \, d\tau$$
$$= 1,$$

the operator J^{-1} exists by the Banach lemma on invertible operators. ∎

Observe that the previous r, which satisfies (4.12), exists if $\omega(R) < \dfrac{1}{2\beta \int_{1/2}^1 h(t) \, dt}$, since ω is a nondecreasing function. Moreover, r is unique. If this condition is not satisfied, r does not exist.

From (4.13), it is easy to see that the uniqueness of the solution is guaranteed in $B(x_0, R)$ if $\omega(R) = \frac{1}{\beta}$.

Next, we analyze the R-order of convergence of Newton's method when conditions (Q1)-(Q2) are satisfied with $h(t) = t^d$, $t \in [0, 1]$ and $d \in [0, 1]$, so that $\omega(tz) \leq t^d \omega(z)$, for $z > 0$. As a consequence, the R-order of convergence is at least $1 + d$. In addition, we observe that the R-order of convergence is at least two if $d = 1$, that is the case in which F' is Lipschitz continuous in Ω.

Note that we cannot apply Ostrowski's technique to see the R-order of convergence of Newton's method under the general conditions given by (Q1)-(Q2). We then introduce a new technique which is based on proving a system of recurrence relations and leads us to obtain a priori error bounds [22].

From (Q1)-(Q2) with $h(t) = t^d$, $t \in [0,1]$ and $d \in [0,1]$, we consider $a_0 = \beta w(\eta)$. Observe that if $x_1 \in \Omega$ and $a_0 < 1$, we have

$$\|I - \Gamma_0 F'(x_1)\| \leq \|\Gamma_0\| \|F'(x_0) - F'(x_1)\| \leq \beta w(\eta) = a_0 < 1.$$

By the Banach lemma on invertible operators, there exists $\Gamma_1 = [F'(x_1)]^{-1}$ and $\|\Gamma_1\| \leq f(a_0)\|\Gamma_0\|$, where f is defined in (4.7). Therefore, x_1 is well defined.

Besides, from Taylor's series and $\{x_n\}$, it follows

$$\|F(x_1)\| = \left\| \int_0^1 (F'(x_0 + \tau(x_1 - x_0)) - F'(x_0))(x_1 - x_0) \, d\tau \right\|$$
$$\leq \left(\int_0^1 w(\tau\eta) \, d\tau \right) \|x_1 - x_0\|$$
$$\leq \frac{w(\eta)}{1 + d} \|x_1 - x_0\|,$$

so that

$$\|x_2 - x_1\| \leq \|\Gamma_1\| \|F(x_1)\| \leq \frac{a_0}{1 + d} f(a_0) \|x_1 - x_0\|,$$
$$\|x_2 - x_0\| \leq \|x_2 - x_1\| + \|x_1 - x_0\| \leq \left(1 + \frac{a_0}{1 + d} f(a_0) \right) \|x_1 - x_0\|,$$
$$\|\Gamma_1\| w(\|x_2 - x_1\|) \leq \beta f(a_0) \left(\frac{a_0}{1 + d} f(a_0) \right)^d w(\|x_1 - x_0\|) \leq \frac{a_0^{1+d}}{(1 + d)^d} f(a_0)^{1+d},$$

provided that $a_0 \leq \dfrac{1 + d}{2 + d}$, since f is increasing in $(0, 1)$.

Note that we can do

$$\frac{a_0^{1+d}}{(1 + d)^d} f(a_0)^{1+d} = a_1$$

and define then the following real sequence:

$$a_n = \frac{a_{n-1}^{1+d}}{(1 + d)^d} f(a_{n-1})^{1+d}, \quad n \geq 1,$$

that satisfies the properties of the following lemma, whose proof is immediate.

Lemma 4.12. *If* $a_0 = \beta w(\eta)$ *satisfies*

$$a_0 \leq \frac{1 + d}{2 + d} \quad \text{and} \quad a_0^d < (1 + d)^d (1 - a_0)^{1+d}, \tag{4.14}$$

then

  (a) the sequence $\{a_n\}$ *is decreasing,*

(b) $a_n \leq \dfrac{1+d}{2+d}$, for all $n \geq 0$.

Since our aim is to obtain a priori error bounds for Newton's method when it converges to a solution x^* of the equation $F(x) = 0$, we first present a system of recurrence relations in the following lemma and some properties of the real sequence $\{a_n\}$ later in Lemma 4.14.

Lemma 4.13. *If the conditions given in (4.14) are satisfied, the following four items are true for all $n \geq 1$:*

(i_n) *There exists $\Gamma_n = [F'(x_n)]^{-1}$ and $\|\Gamma_n\| \leq f(a_{n-1})\|\Gamma_{n-1}\|$,*

(ii_n) $\|x_{n+1} - x_n\| \leq \dfrac{a_{n-1}}{1+d} f(a_{n-1})\|x_n - x_{n-1}\|$,

(iii_n) $\|\Gamma_n\|\omega(\|x_{n+1} - x_n\|) \leq a_n$,

(iv_n) $\|x_{n+1} - x_0\| \leq \dfrac{1 - \Delta^{n+1}}{1 - \Delta}\|x_1 - x_0\| < R\eta$, *where* $\Delta = \dfrac{a_0}{1+d} f(a_0)$ *and* $R = \dfrac{1}{1 - \Delta}$.

From a similar way to the above mentioned for $n = 1$ and using induction, the proof of the previous lemma follows.

Lemma 4.14. *Let f be the scalar function given by (4.7) and $\gamma = \frac{a_1}{a_0}$. If (4.14) is satisfied, then*

(a) $f(\gamma x) < f(x)$, *for $\gamma \in (0, 1)$ and $x \in (0, 1)$,*

(b) $a_n < \gamma^{(1+d)^{n-1}} a_{n-1}$ *and* $a_n < \gamma^{\frac{(1+d)^n - 1}{d}} a_0$, *for all $n \geq 2$.*

Proof. Item (a) is obvious. The proof of item (b) follows by invoking induction hypothesis. As $a_1 = \gamma a_0$ and $\gamma < 1$, then

$$a_2 = \frac{a_1^{1+d}}{(1+d)^d} f(a_1)^{1+d} < \frac{(\gamma a_0)^{1+d}}{(1+d)^d} f(a_0)^{1+d} = \gamma^{1+d} a_1.$$

Now, we suppose $a_{n-1} < \gamma^{(1+d)^{n-2}} a_{n-2} < \gamma^{\frac{(1+d)^{n-1} - 1}{d}} a_0$. As a consequence,

$$a_n = \frac{a_{n-1}^{1+d}}{(1+d)^d} f(a_{n-1})^{1+d}$$

$$< \gamma^{(1+d)^{n-1}} \frac{a_{n-2}^{1+d}}{(1+d)^d} f(a_{n-2})^{1+d}$$

$$= \gamma^{(1+d)^{n-1}} a_{n-1}$$

$$< \gamma^{(1+d)^{n-1}} \gamma^{(1+d)^{n-2}} a_{n-2}$$

$$< \cdots < \gamma^{\frac{(1+d)^n - 1}{d}} a_0.$$

The proof is completed by mathematical induction on n. $\blacksquare$

After that, we obtain the following a priori error bounds and the R-order of convergence.

Theorem 4.15. *Let $F : \Omega \subseteq X \longrightarrow Y$ be a continuously Fréchet differentiable operator defined on a non-empty open convex domain Ω of a Banach space X with values in a Banach space Y. Suppose that (Q1)-(Q2) are satisfied with $h(t) = t^d$, $t \in [0,1]$ and $d \in [0,1]$. Suppose also that $a_0 = \beta\omega(\eta)$ satisfies (4.14) and $B(x_0, R) \subset \Omega$, where $R = \frac{1}{1-\Delta}$ and $\Delta = \frac{a_0}{1+d}f(a_0)$. Then, we have the following a priori error estimates:*

$$\|x^* - x_n\| \le \left(\gamma^{\frac{(1+d)^n-1}{d^2}}\right) \frac{\Theta^n}{1 - \gamma^{\frac{(1+d)^n}{d}}\Theta}\eta, \quad n \ge 0, \tag{4.15}$$

where $\gamma = \frac{a_1}{a_0}$ and $\Theta = \frac{\Delta}{\gamma^{1/d}}$. Moreover, Newton's sequence is of R-order of convergence at least $1 + d$.

Proof. For $m \ge 1$, we have

$$\|x_{n+m} - x_n\| \le \sum_{j=n}^{n+m-1} \|x_{j+1} - x_j\| \le \|x_1 - x_0\| \sum_{j=n-1}^{n+m-2} \left(\prod_{i=0}^{j}\left(\frac{a_i}{1+d}f(a_i)\right)\right)$$

as a consequence of recurrence relation (ii_j) of Lemma 4.13. From

$$\prod_{i=0}^{j}\frac{a_i}{1+d}f(a_i) = \Delta\,\Theta\,\gamma^{\frac{1+d}{d^2}((1+d)^j-1)}$$

and

$$\gamma^{\frac{(1+d)^{n+i}-1}{d^2}} = \gamma^{\frac{(1+d)^n-1}{d^2}}\gamma^{\frac{(1+d)^n}{d^2}((1+d)^i-1)} \le \gamma^{\frac{(1+d)^n-1}{d^2}}\gamma^{\frac{(1+d)^n}{d}i},$$

it follows

$$\|x_{n+m} - x_n\| \le \|x_1 - x_0\| \sum_{j=n-1}^{n+m-2}\left(\Delta\,\Theta^j\,\gamma^{\frac{1+d}{d^2}((1+d)^j-1)}\right)$$

$$= \|x_1 - x_0\| \sum_{i=0}^{m-1}\left(\Theta^{i+n}\,\gamma^{1/d}\,\gamma^{\frac{1+d}{d^2}((1+d)^{i+n-1}-1)}\right)$$

$$\le \|x_1 - x_0\|\,\Theta^n\,\gamma^{\frac{(1+d)^n-1}{d^2}} \sum_{i=0}^{m-1}\left(\Theta\,\gamma^{\frac{(1+d)^n}{d}}\right)^i.$$

Therefore,

$$\|x_{n+m} - x_n\| < \frac{1 - \left(\Theta\,\gamma^{\frac{(1+d)^n}{d}}\right)^m}{1 - \Theta\,\gamma^{\frac{(1+d)^n}{d}}}\,\Theta^n\,\eta\,\gamma^{\frac{(1+d)^n-1}{d^2}}. \tag{4.16}$$

By letting $m \to +\infty$ in (4.16), we obtain (4.15).

Now, from (4.15), it follows that

$$\|x^* - x_n\| \le \frac{\eta}{\gamma^{1/d^2}(1 - \Theta)}\left(\gamma^{1/d^2}\right)^{(1+d)^n}, \quad n \ge 0,$$

and the R-order of convergence of the Newton sequence is therefore at least $1 + d$. ∎

Remark 4.16. First, if F' is Lipschitz continuous in Ω, then $\omega(z) = Mz$, $M \geq 0$, $\omega(tz) \leq t\omega(z)$ and the R-order of convergence is at least two, as we already know. Second, if F' is Hölder continuous in Ω, then $\omega(z) = Kz^p$, $K \geq 0$, $p \in [0,1]$, $\omega(tz) \leq t^p\omega(z)$ and the R-order of convergence is at least $1 + p$. Third, if F' is such that

$$\|F'(x) - F'(y)\| \leq \sum_{i=1}^{j} K_i \|x - y\|^{d_i}, \quad d_i \in [0,1], \quad x, y \in \Omega,$$

we can consider $\omega(z) = \sum_{i=1}^{j} K_i z^{d_i}$, that satisfies

$$\omega(tz) = \sum_{i=1}^{j} K_i t^{d_i} z^{d_i} \leq t^d \omega(z),$$

where $d = \max_{i=1,2,\ldots,j}\{d_i\}$, $d_i \in [0,1]$, for $i = 1, 2, \ldots, j$, and $t \in [0,1]$, so that the R-order of convergence is at least $1 + d$.

4.2.3 Application

In this section, we provide some results of existence and uniqueness of solution for nonlinear integral equations of Hammerstein-type (1.36),

$$x(s) = u(s) + \int_a^b \mathcal{K}(s,t)\mathcal{H}(x(t))\, dt, \quad s \in [a,b],$$

where

$$\mathcal{H}(x(t)) = \lambda_1\, x(t)^{1+p} + \lambda_2\, x(t)^q, \quad p \in [0,1], \quad q \in \mathbb{N}, \quad q \geq 2, \quad \lambda_1, \lambda_2 \in \mathbb{R}, \tag{4.17}$$

u is a continuous function such that $u(s) > 0$ and the kernel $\mathcal{K}(s,t)$ is continuous and nonnegative in $[a,b] \times [a,b]$.

Remark 4.17. Note that if the kernel $\mathcal{K}(s,t)$ is the Green function given in (1.50), then the integral equation given by (1.36)-(4.17) is equivalent to the following boundary value problem:

$$\begin{cases} x''(s) = -\lambda_1 x(s)^{1+p} - \lambda_2 x(s)^q, \\ x(a) = u(a), \quad x(b) = u(b). \end{cases}$$

Existence and uniqueness of solution

Observe that solving equation (1.36)-(4.17) is equivalent to solving $\mathcal{F}(x) = 0$, where $\mathcal{F} : \Omega \subseteq \mathcal{C}([a,b]) \longrightarrow \mathcal{C}([a,b])$ and

$$[\mathcal{F}(x)](s) = x(s) - u(s) - \int_a^b \mathcal{K}(s,t)\left(\lambda_1 x(t)^{1+p} + \lambda_2 x(t)^q\right) dt, \tag{4.18}$$

where $p \in (0,1]$, $q \in \mathbb{N}$, $q \geq 2$ and $\lambda_1, \lambda_2 \in \mathbb{R}$. We then apply the study of the last section to obtain different results on the existence and uniqueness of solution of equation (1.36)-(4.17).

First of all, we determine the domain Ω. For this, $x(s)$ must be such that $x(s) \geq 0$, $s \in [a,b]$, so that the existence of $x(s)^{1+p}$ is guaranteed for each $p \in [0,1]$. Moreover, as F' is ω-Lipschitz, F' satisfies (Q2) and this fact depends on Ω.

The first derivative of operator (4.18) is

$$[\mathcal{F}'(x)y](s) = y(s) - \int_a^b \mathcal{K}(s,t)\left((1+p)\lambda_1 x(t)^p + q\lambda_2 x(t)^{q-1}\right)y(t)\,dt \qquad (4.19)$$

and

$$[(\mathcal{F}'(x)-\mathcal{F}'(y))z](s) = -\int_a^b \mathcal{K}(s,t)\left((1+p)\lambda_1\left(x(t)^p - y(t)^p\right) + q\lambda_2\left(x(t)^{q-1} - y(t)^{q-1}\right)\right)z(t)\,dt.$$

If $q = 2$, we have

$$\|\mathcal{F}'(x) - \mathcal{F}'(y)\| \le S\left((1+p)|\lambda_1|\|x - y\|^p + 2|\lambda_2|\|x - y\|\right),$$

where $S = \left\|\int_a^b \mathcal{K}(s,t)dt\right\|$. If $q \ge 3$, we then have

$$\|\mathcal{F}'(x) - \mathcal{F}'(y)\| \le S\Big((1+p)|\lambda_1|\|x - y\|^p$$
$$+ q|\lambda_2|\left(\|x\|^{q-2} + \|x\|^{q-3}\|y\| + \cdots + \|x\|\|y\|^{q-3} + \|y\|^{q-2}\right)\|x - y\|\Big)$$
$$= S\left((1+p)|\lambda_1|\|x - y\|^p + q|\lambda_2|\left(\sum_{i=0}^{q-2}\|x\|^{q-2-i}\|y\|^i\right)\|x - y\|\right).$$

As a consequence of the last, to obtain the function ω, we have to bound $\|x\|$ and $\|y\|$, for what we need to fix the domain Ω. Besides, as a solution $x^*(s)$ of the equation must be contained in Ω, a previous location of $x^*(s)$ is usually done. For this, from (1.36)-(4.17), it follows

$$\|x^*(s)\| \le U + S\|\mathcal{H}(x^*(t))\| \le U + S\left(|\lambda_1|\|x^*(t)\|^{1+p} + |\lambda_2|\|x^*(t)\|^q\right)$$

and

$$0 \le U + S\left(|\lambda_1|\|x^*(s)\|^{1+p} + |\lambda_2|\|x^*(s)\|^q\right) - \|x^*(s)\|, \qquad (4.20)$$

where $U = \|u(s)\|$. So, we consider the scalar equation deduced from the last expression and given by

$$\chi(t) = U + S\left(|\lambda_1|\,t^{1+p} + |\lambda_2|\,t^q\right) - t = 0 \qquad (4.21)$$

and suppose that equation (4.21) has at least one positive real solution. We then denote the smallest positive real solution by ρ_1.

Obviously, if the last is true, condition (4.20) is satisfied, provided that $\|x^*(s)\| < \rho_1$, so that if integral equation (1.36)-(4.17) has a solution $x^*(s) \in B(0,\rho_1)$, we can choose

$$\Omega = \{x \in \mathcal{C}([a,b]): \; x(s) > 0, \; \|x(s)\| < \rho, \; s \in [a,b]\}, \qquad (4.22)$$

for some $\rho > \rho_1$, since Ω is an open domain. As a consequence,

$$\omega(z) = S\left((1+p)|\lambda_1|z^p + q(q-1)|\lambda_2|\rho^{q-2}z\right) \qquad (4.23)$$

and, in addition, $\omega(tz) \le h(t)\omega(z)$ with $h(t) = t^p$, for all $q \in \mathbb{N}$ and $q \ge 2$.

Note that, depending on what the exponents of $x(t)$ are into the integral equation given by (1.36)-(4.17), the condition $x(s) > 0$ appearing in the domain Ω can omit.

On the contrary, if we cannot guarantee the existence of a solution of integral equation (1.36)-(4.17) in a ball $B(0,r)$, we cannot find a domain Ω containing a solution $x^*(s)$ of (1.36)-(4.17) and where condition (4.20) is satisfied. In these cases, as we have see in Chapters 2 and 3, we can consider higher order derivatives of F that are ω-bounded in Ω and for which it is not necessary to fix a priori the domain Ω (see Sections 2.2 and 3.2).

We now calculating the parameters β and η that appear in the study. So, if $x_0(s)$ is fixed, then

$$\|I - \mathcal{F}'(x_0)\| \leq S\left((1+p)|\lambda_1|\|x_0^p\| + q|\lambda_2|\|x_0^{q-1}\|\right).$$

By the Banach lemma on invertible operators, if $S\left((1+p)|\lambda_1|\|x_0^p\| + q|\lambda_2|\|x_0^{q-1}\|\right) < 1$, we obtain

$$\|[\mathcal{F}'(x_0)]^{-1}\| \leq \frac{1}{1 - S\left((1+p)|\lambda_1|\|x_0^p\| + q|\lambda_2|\|x_0^{q-1}\|\right)} = \beta. \tag{4.24}$$

From the definition of the operator $\mathcal{F}$, we have $\|\mathcal{F}(x_0)\| \leq \|x_0 - u\| + S\left(|\lambda_1|\left\|x_0^{1+p}\right\| + |\lambda_2|\|x_0^q\|\right)$, and therefore

$$\|[\mathcal{F}'(x_0)]^{-1}\mathcal{F}(x_0)\| \leq \frac{\|x_0 - u\| + S\left(|\lambda_1|\left\|x_0^{1+p}\right\| + |\lambda_2|\|x_0^q\|\right)}{1 - S\left((1+p)|\lambda_1|\|x_0^p\| + q|\lambda_2|\|x_0^{q-1}\|\right)} = \eta. \tag{4.25}$$

Once the parameters β and η are calculated and the function ω is known, we can already establish the following result on the existence of solution of equation (1.36)-(4.17) from Theorem 4.9.

Theorem 4.18. *Let $\mathcal{F}$ be the operator defined in (4.18) with Ω given in (4.22) and $x_0 \in \Omega$ such that $S\left((1+p)|\lambda_1|\|x_0^p\| + q|\lambda_2|\|x_0^{q-1}\|\right) < 1$, where $S = \left\|\int_a^b \mathcal{K}(s,t)dt\right\|$. If*

$$\beta\omega(\eta) \leq \frac{1+p}{2+p} \quad and \quad \beta\omega(\eta) < 1 - \left(\frac{\beta\omega(\eta)}{(1+p)(1-\beta\omega(\eta))}\right)^p,$$

with β given in (4.24), η in (4.25) and ω in (4.23), and $B(x_0, R) \subset \Omega$, where $R = \frac{(1+p)(1-\beta\omega(\eta))}{(1+p)-(2+p)\beta\omega(\eta)}\eta$, then a solution of (1.36)-(4.17) exists at least in $\overline{B(x_0, R)}$. Moreover, this solution is unique in $\Omega_0 = B(x_0, r) \cap \Omega$, where r is the smallest positive real root of the equation

$$2\beta\omega(R+t)(2^{1+p} - 1) = (1+p)2^{1+p}.$$

Observe that Newton's sequence $\{x_n\}$ is also convergent if $\beta\omega(\eta) = 1 - \left(\frac{\beta\omega(\eta)}{(1+p)(1-\beta\omega(\eta))}\right)^p <$ $\frac{1+p}{2+p}$, see Remark 4.10.

Note also that the bound given for $\|\mathcal{F}(x_0)\|$ can be improved when the kernel $\mathcal{K}(s,t)$ and the function u are fixed.

Remark 4.19. The cases $p = 0$ and $q = 1$ correspond to situations in which F' satisfies a Lipschitz condition or Hölder condition in some domain Ω, respectively. In both cases, we can do an analogous study for these types of integral equations from the respective semilocal convergence results given later in Section 4.2.4.

Example 4.20. If we consider the particular case of equation (1.36)-(4.17) given by

$$x(s) = 1 + \int_0^1 G(s,t)\left(x(t)^{\frac{3}{2}} + \frac{1}{2}x(t)^2\right) dt, \quad s \in [0,1], \tag{4.26}$$

and choose

$$\Omega = \{x \in \mathcal{C}([0,1]) : x(s) > 0, \ s \in [0,1]\},$$

instead of a domain of type (4.22), since we do not need to locate previously a solution of equation (4.26), and $x_0(s) = 1$, we have

$$u(s) = 1, \quad p = \frac{1}{2}, \quad q = 2, \quad \lambda_1 = 1, \quad \lambda_2 = \frac{1}{2},$$

$$\beta = \frac{16}{11}, \quad \eta = \frac{3}{11}, \quad w(z) = \frac{1}{16}\left(3\sqrt{z} + 2z\right), \quad h(t) = \sqrt{t}, \quad R = 0.3240\ldots$$

Thus, as $S\left((1+p)|\lambda_1|\|x_0^p\| + q|\lambda_2|\|x_0^{q-1}\|\right) = \frac{5}{16} < 1$, $\beta w(\eta) = 0.1920\ldots \leq \frac{1+p}{2+p} = \frac{3}{5} = 0.6$, $\beta w(\eta) = 0.1920\ldots < 1 - \left(\frac{\beta w(\eta)}{(1+p)(1-\beta w(\eta))}\right)^p = 0.6019\ldots$ and $B(x_0, R) = B(1, 0.3240\ldots) \subset \Omega$, the hypotheses of Theorem 4.18 are satisfied. Then (4.26) has a solution x^* in the region $\{v \in \Omega : \|v - 1\| \leq 0.3240\ldots\}$, which is unique in $\{v \in \Omega : \|v - 1\| < 3.2293\ldots\} \cap \Omega$.

Next, we approximate numerically a solution of integral equation (4.26). For this, we use a process of discretization to transform equation (4.26) into a finite dimensional problem by approximating the integral of (4.26) by a Gauss-Legendre quadrature formula with 8 nodes:

$$\int_a^b \varphi(t)\, dt \simeq \sum_{i=1}^{8} w_i \varphi(t_i),$$

where the nodes t_i and the weights w_i are determined.

If we denote the approximations of $x(t_i)$ by x_i, with $i = 1, 2, \ldots, 8$, then equation (4.26) is equivalent to the following system of nonlinear equations:

$$x_i = 1 + \sum_{j=1}^{8} a_{ij}\left(x_j^{3/2} + \frac{1}{2}x_j^2\right), \quad j = 1, 2, \ldots, 8, \tag{4.27}$$

where

$$a_{ij} = w_j G(t_i, t_j) = \begin{cases} w_j(1 - t_i)t_j, & j \leq i, \\ w_j t_i(1 - t_j), & j > i. \end{cases}$$

Now, we write system (4.27) as

$$\mathbb{F}(\mathbf{x}) \equiv \mathbf{x} - \mathbf{1} - A\,\hat{\mathbf{x}} = 0, \tag{4.28}$$

where $\mathbb{F} : \mathbb{R}^8 \longrightarrow \mathbb{R}^8$ and

$$\mathbf{x} = (x_1, x_2, \ldots, x_8)^T, \quad \mathbf{1} = (1, 1, \ldots, 1)^T, \quad A = (a_{ij})_{i,j=1}^8,$$

$$\hat{\mathbf{x}} = \left(x_1^{3/2} + \frac{1}{2}x_1^2, x_2^{3/2} + \frac{1}{2}x_2^2, \ldots, x_8^{3/2} + \frac{1}{2}x_8^2\right)^T.$$

Besides,

$$\mathbb{F}'(\mathbf{x}) = I - A\left(\frac{3}{2}\mathrm{diag}\left\{x_1^{1/2}, x_2^{1/2}, \ldots, x_8^{1/2}\right\} + \mathrm{diag}\{x_1, x_2, \ldots, x_8\}\right),$$

where I denotes the identity matrix. Since $x_0(s) = 1$ has been chosen as starting point for the theoretical study, a reasonable choice of initial approximation for Newton's method seems to be the vector $\mathbf{x_0} = \mathbf{1}$. After four iterations, we obtain the numerical approximation to the solution $\mathbf{x}^* = (x_1^*, x_2^*, \ldots, x_8^*)^T$ shown in Table 4.1.

i	x_i^*	i	x_i^*
1	1.01960381...	5	1.25935157...
2	1.09396463...	6	1.19091775...
3	1.19091775...	7	1.09396463...
4	1.25935157...	8	1.01960381...

Table 4.1: Numerical solution $\mathbf{x}^*$ of system (4.28)

n	$\|\mathbf{x}^* - \mathbf{x_n}\|$	$\|F(\mathbf{x_n})\|$
0	$2.5935\ldots \times 10^{-1}$	$1.8533\ldots \times 10^{-1}$
1	$7.4625\ldots \times 10^{-3}$	$5.2360\ldots \times 10^{-3}$
2	$6.1224\ldots \times 10^{-6}$	$4.2986\ldots \times 10^{-6}$
3	$4.1031\ldots \times 10^{-12}$	$2.8814\ldots \times 10^{-12}$

Table 4.2: Absolute errors and $\{\|F(\mathbf{x_n})\|\}$

In Table 4.2 we show the errors $\|\mathbf{x}^* - \mathbf{x_n}\|$ and the sequence $\{\|\mathbb{F}(\mathbf{x_n})\|\}$. From the last, we notice that the vector shown in Table 4.1 is a good approximation of the solution of system (4.28)

We now interpolate the points of Table 4.1. Taking into account that the solution of (4.26) satisfies $x(0) = x(1) = 1$, the approximation of the numerical solution drawn in Figure 4.1 is obtained. Notice that the interpolated approximation lies within the existence domain of the solutions obtained previously: $\{\nu \in \mathcal{C}([0,1]) : \|\nu - 1\| \leq 0.3240\ldots\}$.

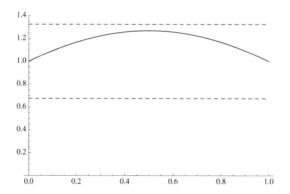

Figure 4.1: Graph (the solid line) of the approximated solution $\mathbf{x}^*$ of system (4.28)

4.2.4 Particular cases

We now comment the two main particular cases of operators with ω-Lipschitz first derivate. In particular, those in which the first derivative of the operator involved is Lipschitz continuous in Ω ($\omega(z) = Mz$ and $h(t) = t$ in (P5)) and Hölder continuous in Ω ($\omega(z) = Kz^p$ and $h(t) = t^p$ with $p \in [0, 1]$ in (P5)).

4.2.4.1   Operators with Lipschitz first-derivative

If $\omega(z) = Mz$ and $h(t) = t$ in (P5), F' is Lipschitz continuous in Ω and the conditions required in Theorems 4.9 and 4.11 are reduced to those appearing in Ortega's version of the Newton-Kantorovich theorem (Theorem 4.6), so that the semilocal convergence of Newton's method and uniqueness of solution are guaranteed under the same hypotheses of the Newton-Kantorovich theorem, as we can see in the next result [24].

Theorem 4.21. *Let $F : \Omega \subseteq X \longrightarrow Y$ be a continuously Fréchet differentiable operator defined on a non-empty open convex domain Ω of a Banach space X with values in a Banach space Y. Suppose that conditions (Q1)-(P2) are satisfied. Suppose also that $b_0 = M\beta\eta \leq \frac{1}{2}$ and $B(x_0, R) \subset \Omega$, where $R = \frac{2(1-b_0)}{2-3b_0}\eta$. Then, Newton's sequence $\{x_n\}$ converges to a solution x^* of $F(x) = 0$ starting at x_0. Moreover, $x_n, x^* \in \overline{B(x_0, R)}$ and x^* is unique in $B(x_0, r) \cap \Omega$, where $r = \frac{2(b_0^2 - 4b_0 + 2)}{b_0(2-3b_0)}\eta$.*

Observe that the main convergence condition, $b_0 = M\beta\eta \leq \frac{1}{2}$, that guarantees the semilocal convergence of Newton's method, is the same as that required in the variant of the Newton-Kantorovich theorem given by Ortega (Theorem 4.6).

Note that the domains of existence and uniqueness of solution that are obtained from the technique of recurrence relations, represented by Theorem 4.21, seem to be better than those obtained from the method of majorizing sequences, represented respectively by Theorems 4.6 and 4.5, as we can see in the following example.

Example 4.22. If we consider the nonlinear integral equation of type (1.36)-(4.17) given by

$$x(s) = 1 + \int_0^1 G(s,t)(1 + x(t))x(t)\, dt, \quad s \in [0,1], \tag{4.29}$$

we have that the operator $\mathcal{F}'$ corresponding to (4.29) and given by (4.19) is

$$[\mathcal{F}'(x)y](s) = y(s) - \int_0^1 G(s,t)(1 + 2x(t))y(t)\, dt,$$

which is Lipschitz continuous in $\Omega = \mathcal{C}([0,1])$ with Lipschitz constant $M = \frac{1}{4}$.

If we choose $x_0(s) = 1$, then $\beta = 2$, $\eta = \frac{1}{2}$, $b_0 = M\beta\eta = \frac{1}{4}$ and $B(x_0, R) \subset \Omega$ with $R = 0.6$, so that the hypotheses of Theorem 4.21 are satisfied and Newton's method is then convergent to a solution x^*. Note that the domains of existence and uniqueness of solution are respectively:

$$\{\nu \in \mathcal{C}([0,1]) : \|\nu(s) - x_0(s)\| \leq 0.5857\ldots\} \quad \text{and} \quad \{\nu \in \mathcal{C}([0,1]) : \|\nu(s) - x_0(s)\| < 3.4\},$$

which are better than those obtained from Ortega's version of the Newton-Kantorovich theorem (Theorem 4.6), where the method of majorizing sequences is used to guarantee the semilocal convergence of Newton's method, that are respectively:

$$\{\nu \in \mathcal{C}([0,1]) : \|\nu(s) - x_0(s)\| \leq 0.6\} \quad \text{and} \quad \{\nu \in \mathcal{C}([0,1]) : \|\nu(s) - x_0(s)\| < 2.0666\ldots\}.$$

4.2.4.2 Operators with Hölder first-derivative

If $\omega(z) = Kz^p$ and $h(t) = t^p$ with $p \in [0,1]$, then Theorem 4.9 and Theorem 4.11 are reduced to the following result [43].

Theorem 4.23. *Let* $F : \Omega \subseteq X \longrightarrow Y$ *be a continuously Fréchet differentiable operator defined on a non-empty open convex domain* Ω *of a Banach space* X *with values in a Banach space* Y. *Suppose that conditions (Q1)-(P3) are satisfied. Suppose also that* $b_0 = K\beta\eta^p \leq \xi$, *where* ξ *is the unique zero of the function*

$$\phi(t;p) = (1+p)^p(1-t)^{1+p} - t^p \tag{4.30}$$

in the interval $(0, 1/2]$ *and* $B(x_0, R) \subset \Omega$, *where* $R = \frac{(1+p)(1-b_0)}{(1+p)-(2+p)b_0}\eta$. *Then, Newton's sequence* $\{x_n\}$ *converges to a solution* x^* *of* $F(x) = 0$ *starting at* x_0. *Moreover,* $x_n, x^* \in \overline{B(x_0, R)}$ *and* x^* *is unique in* $B(x_0, r) \cap \Omega$, *where* r *is the smallest positive real solution of the equation* $K\beta(r^{1+p} - R^{1+p}) + (1+p)(R-r) = 0$.

Observe that in the previous result we have not included the value $p = 0$ in condition (P3). For this value, condition (P3) is reduced to

$$\|F'(x) - F'(y)\| \leq K, \quad x, y \in \Omega, \tag{4.31}$$

and the semilocal convergence result for Newton's method is then the following (see [43]):

Theorem 4.24. *Let* $F : \Omega \subseteq X \longrightarrow Y$ *be a continuously Fréchet differentiable operator defined on a non-empty open convex domain* Ω *of a Banach space* X *with values in a Banach space* Y. *Suppose that conditions (Q1)-(4.31) are satisfied. If* $b_0 = K\beta \in (0, 1/2)$ *and* $B(x_0, R) \subset \Omega$ *with* $R = \frac{1-b_0}{1-2b_0}\eta$, *then Newton's sequence* $\{x_n\}$ *converges to a solution* x^* *of* $F(x) = 0$ *starting at* x_0. *Moreover,* $x_n, x^* \in \overline{B(x_0, R)}$.

Now, we consider an example, where the first Fréchet derivative of the operator involved is Hölder continuous in the domain, and then use Theorem 4.23 to guarantee the semilocal convergence of Newton's method.

Example 4.25. We consider the nonlinear integral equation of type (1.36)-(4.17) given by

$$x(s) = s^2 + 2\int_0^1 G(s,t)x(t)^{5/3}\,dt, \quad s \in [0,1]. \tag{4.32}$$

In this case, the operator $\mathcal{F}'$ corresponding to (4.43) and given by (4.19) is

$$[\mathcal{F}'(x)y](s) = y(s) - \frac{10}{3}\int_0^1 G(s,t)x(t)^{2/3}y(t)\,dt,$$

which is Hölder continuous in $\Omega = \mathcal{C}([0,1])$, where Hölder's constants are $K = \frac{5}{12}$ and $p = \frac{2}{3}$. Then, if we choose $x_0(s) = s^2$, then $\beta = \frac{12}{7}$, $\eta = \frac{3}{7}$ and $b_0 = K\beta\eta^p = 0.4060\ldots$, so that the hypotheses of Theorem 4.23 are satisfied, since $b_0 \leq \xi = 0.42246\ldots$, where ξ is the unique positive real zero of function (4.30), $\phi(t;p) = \left(\frac{5}{3}\right)^{2/3}(1-t)^{5/3} - t^{2/3} = 0$ in the interval $(0, 1/2]$ and $B(x_0, R) \subset \Omega$ with $R = 0.7265\ldots$ are satisfied. Therefore, the semilocal convergence of Newton's method is guaranteed from Theorem 4.23 and the domains of existence and uniqueness of solution are respectively:

$$\{v \in \mathcal{C}([0,1]) : \|v(s) - x_0(s)\| \leq 0.7265\ldots\} \text{ and } \{v \in \mathcal{C}([0,1]) : \|v(s) - x_0(s)\| < 2.6506\ldots\}.$$

4.3 Operators with center ω-Lipschitz first-derivative

The aim of this section is to prove the semilocal convergence of Newton's method under milder conditions than the previous ones. In particular, the ω-Lipschitz condition on F' required in Section 4.2 is relaxed. For this, we require a center ω-Lipschitz condition on F' instead of a ω-Lipschitz condition. So, we fix one of the two points of the ω-Lipschitz condition, so that this point is the starting point x_0 of Newton's method and the condition is then satisfied for any $x \in \Omega$ and x_0 instead of any two points $x, y \in \Omega$.

Obviously, if (P2) is satisfied, we have

$$\|F'(x) - F'(x_0)\| \leq M_0 \|x - x_0\|, \quad x \in \Omega, \tag{4.33}$$

with $M_0 \leq M$ and F' is then center Lipschitz continuous at x_0. In addition, condition (4.33) can be satisfied, but condition (P2) cannot, as we can see in the following.

If we consider a nonlinear integral equation of the Hammerstein-type given by (1.36)-(4.17) with $\lambda_2 = 0$ and $p = \frac{1}{n}$ with $n \in \mathbb{N}$; i.e.:

$$x(s) = u(s) + \lambda_1 \int_a^b \mathcal{K}(s,t)\, x(t)^{1+\frac{1}{n}}\, dt, \quad s \in [a,b], \quad n \in \mathbb{N}, \quad \lambda_1 \in \mathbb{R}, \tag{4.34}$$

then solving equation (4.34) is equivalent to solving the equation $\mathcal{F}_1(x) = 0$, where $\mathcal{F}_1 : \Omega_1 \subseteq \mathcal{C}([a,b]) \longrightarrow \mathcal{C}([a,b])$ and

$$[\mathcal{F}_1(x)](s) = x(s) - u(s) - \lambda_1 \int_a^b \mathcal{K}(s,t)\, x(t)^{1+\frac{1}{n}}\, dt, \quad s \in [a,b], \quad n \in \mathbb{N}, \quad \lambda_1 \in \mathbb{R}.$$

In addition,

$$[\mathcal{F}_1'(x)y](s) = y(s) - \lambda_1 \left(1 + \frac{1}{n}\right) \int_a^b \mathcal{K}(s,t)\, x(t)^{\frac{1}{n}}\, y(t)\, dt. \tag{4.35}$$

If we now consider $\Omega_1 = \{x \in \mathcal{C}([a,b]) : x(s) > 0, \ s \in [a,b]\}$ if n is even, or $\Omega_1 = \mathcal{C}([a,b])$ if n is odd, and follow a development completely analogous to that seen in Case 3 of Section 2.1.3.6 for integral equation (2.28), where the second derivative of the operator involved is center Lipschitz continuous at the starting point of Newton's method, but it is not in all domain, we see that (4.35) is not Lipschitz continuous in all Ω_1, so that condition (P2) is not satisfied, but is center Lipschitz at $x_0 \in \Omega_1$, since condition (4.33) is satisfied.

On the other hand, taking into account condition (4.33) with $M_0 \leq M$ and the semilocal convergence result that we obtain later for Newton's method under this condition (see Corollary 4.30), it is necessary to satisfy the condition $M_0 \beta \eta \leq \zeta$ with $\zeta < \frac{1}{2}$, so that the convergence condition can be more restrictive than that, $M\beta\eta \leq \frac{1}{2}$, required for the Lipschitz case. Moreover, as $M_0 \leq M$, the condition $M_0 \beta \eta \leq \zeta$ can be satisfied, but not the condition $M\beta\eta \leq \frac{1}{2}$, so that we can find new good starting points x_0 for Newton's method, as we can see in the following.

We now consider a nonlinear integral equation of the Hammerstein-type given by (1.36)-(4.17) with $\lambda_1 = 0$, $q \in \mathbb{N}$ and $q \geq 3$; i.e.:

$$x(s) = u(s) + \lambda_2 \int_a^b \mathcal{K}(s,t)\, x(t)^q\, dt, \quad s \in [a,b], \quad q \in \mathbb{N}, \quad q \geq 3, \quad \lambda_2 \in \mathbb{R}. \tag{4.36}$$

Solving equation (4.36) is equivalent to solving the equation $\mathcal{F}_2(x) = 0$, where

$$\mathcal{F}_2 : \Omega_2 \subseteq \mathcal{C}([a,b]) \longrightarrow \mathcal{C}([a,b]),$$

$$[\mathcal{F}_2(x)](s) = x(s) - u(s) - \lambda_2 \int_a^b \mathcal{K}(s,t)\, x(t)^q\, dt, \quad s \in [a,b], \quad q \in \mathbb{N}, \quad q \geq 3, \quad \lambda \in \mathbb{R}.$$

In this case, we consider

$$\Omega_2 = \{x \in \mathcal{C}([a,b]) : \ \|x(s) - \nu(s)\| < r, \ s \in [a,b]\}$$

and have

$$[\mathcal{F}_2'(x)z](s) = z(s) - \lambda_2 q \int_a^b \mathcal{K}(s,t)\, x(t)^{q-1}\, z(t)\, dt.$$

As a consequence,

$$\|\mathcal{F}_2'(x) - \mathcal{F}_2'(y)\| \leq |\lambda_2| q S \left(\|x\|^{q-2} + \|x\|^{q-3}\|y\| + \cdots + \|x\|\|y\|^{q-3} + \|y\|^{q-2} \right) \|x - y\|$$
$$\leq |\lambda_2| q S(q-1) \left(\|\nu\| + r \right)^{q-2} \|x - y\|,$$

where $S = \left\| \int_a^b \mathcal{K}(s,t)\, dt \right\|$. Therefore, if $x, y \in \Omega_2$, then $\mathcal{F}_2'$ is Lipschitz continuous in Ω_2 and the Lipschitz constant is $M = |\lambda_2| q S(q-1) \left(\|\nu\| + r \right)^{q-2}$, and, if $y = u$ is fixed in Ω_2, then $\mathcal{F}_2'$ is center Lipschitz at u and the Lipschitz constant is $M_0 = |\lambda_2| q S \left(\sum_{i=0}^{q-2} \|u\|^i \left(\|\nu\| + r \right)^{q-2-i} \right)$. Obviously, depending on the values of $\|u\|$, $\|\nu\|$ and r, we have $M_0 \ll M$, thus confirming our intention to modify the domain of starting points, as we see later in Example 4.29.

After that, as we have done with the center conditions for the operator F'' in Section 2.1, we now generalize condition (4.33). So, we investigate if it is possible to weaken condition (P2) by assuming that

$$\|F'(x) - F'(x_0)\| \leq \omega_0(\|x - x_0\|), \quad x \in \Omega, \tag{4.37}$$

where the real function $\omega_0 : [0, +\infty) \longrightarrow \mathbb{R}$ is continuous and nondecreasing and such that $\omega_0(0) = 0$.

First, we notice that (P2) obviously implies (4.37). However, as we have seen before, the reciprocal is not true.

4.3.1   Semilocal convergence

We now suppose that the operator F satisfies the following conditions:

(R1) There exists $\Gamma_0 = [F'(x_0)]^{-1} \in \mathcal{L}(Y,X)$, for some $x_0 \in \Omega$, with $\|\Gamma_0\| \leq \beta$ and $\|\Gamma_0 F(x_0)\| \leq \eta$.

(R2) There exist two continuous and nondecreasing functions $\omega_0 : [0, +\infty) \longrightarrow \mathbb{R}$ and $h_0 : [0,1] \longrightarrow \mathbb{R}$ such that $\omega_0(0) = 0$, $\omega_0(tz) \leq h_0(t)\omega_0(z)$, with $t \in [0,1]$, $z \in [0, +\infty)$ and $\|F'(x) - F'(x_0)\| \leq \omega_0(\|x - x_0\|)$ with $x \in \Omega$.

(R3) There exits at least one positive real solution of the scalar equation

$$g(t) = (1 - (3 - I_{h_0})\beta\omega_0(t))\eta - (1 - 3\beta\omega_0(t))t = 0,$$

where $I_{h_0} = \int_0^1 h_0(\tau)\, d\tau$. We denote the small positive solution of this equation by R.

(R4) $\beta\omega_0(R) < \frac{1}{3}$ and $B(x_0, R) \subset \Omega$.

Our aim in this section is to prove that Newton's method converges to a solution of the equation $F(x) = 0$ under conditions (R1)-(R2)-(R3)-(R4). In addition, we give the domains where the solution is located and is unique. For this, we follow a technique based on recurrence relations which is similar to that developed in Section 4.2. Finally, we illustrate the theoretical results with an application to a nonlinear integral equation of the Hammerstein type. First at all, we give two technical lemmas.

Lemma 4.26. *Suppose (R1)-(R2)-(R3)-(R4) and define* $f(t) = \frac{1}{1-\beta\omega_0(t)}$. *Then,*

$$\left(1 + I_{h_0}\sum_{i=1}^{j} 2^{i-1}(\beta\omega_0(R)f(R))^i\right)\eta < \left(1 + \frac{I_{h_0}\beta\omega_0(R)f(R)}{1 - 2\beta\omega_0(R)f(R)}\right)\eta = \frac{(1 - (3 - I_{h_0})\beta\omega_0(R))\eta}{1 - 3\beta\omega_0(R)} = R.$$

Proof. For the inequality, it suffices to calculate the sum of a geometric series with ratio $2\beta\omega_0(R)f(R)$. This ratio is less than 1 if and only if $\beta\omega_0(R) < \frac{1}{3}$. The equality is satisfied from (R3). ∎

Lemma 4.27. *Under conditions (R1)-(R2)-(R3)-(R4) and the notations of the previous lemma, we have:*

(i_n) *There exists* $\Gamma_n = [F'(x_n)]^{-1}$ *and* $\|\Gamma_n F'(x_0)\| \leq f(R)$, $n \geq 1$.

(ii_n) $\|\Gamma_0 F(x_n)\| \leq 2\beta\omega_0(R)\|x_n - x_{n-1}\|$, $n \geq 2$.

(iii_n) $\|x_{n+1} - x_n\| \leq 2\beta\omega_0(R)f(R)\|x_n - x_{n-1}\|$, $n \geq 2$.

(iv_n) $\|x_{n+1} - x_0\| \leq \left(1 + I_{h_0}\sum_{j=0}^{n-1} 2^j (\beta\omega_0(R)f(R))^{j+1}\right)\eta < R$, $n \geq 1$.

Proof. First, we note that

$$\|I - \Gamma_0 F'(x)\| = \|\Gamma_0\|\|F'(x) - F'(x_0)\| \leq \beta\omega_0(\|x - x_0\|) \leq \beta\omega_0(R) < 1$$

if $x \in B(x_0, R)$. Then, there exists $[F'(x)]^{-1}$ and

$$\|[F'(x)]^{-1}F'(x_0)\| \leq \frac{1}{1 - \beta\omega_0(R)} = f(R).$$

Now, as $\|x_1 - x_0\| \leq \eta < R$, then $x_1 \in B(x_0, R) \subset \Omega$ and $\|\Gamma_1 F'(x_0)\| \leq f(R)$ from condition (R4). Next, from Taylor's series, we obtain

$$\|F(x_1)\| = \left\|\int_{x_0}^{x_1} (F'(x) - F'(x_0))\, dx\right\|$$

$$= \left\|\int_0^1 (F'(x_0 + \tau(x_1 - x_0)) - F'(x_0))(x_1 - x_0)\, d\tau\right\|$$

$$\leq \int_0^1 \beta\omega_0(\tau\|x_1 - x_0\|)\, d\tau \|x_1 - x_0\|$$

$$\leq \int_0^1 h_0(\tau)\, d\tau \beta\omega_0(R)\|x_1 - x_0\|$$

$$= I_{h_0}\beta\omega_0(R)\|x_1 - x_0\|$$

and

$$\|x_2 - x_1\| \leq \|\Gamma_1 F'(x_0)\| \|\Gamma_0\| \|F(x_1)\| \leq I_{h_0}\beta\omega_0(R)f(R)\|x_1 - x_0\|. \tag{4.38}$$

So, from Lemma 4.26, it follows

$$\begin{aligned}
\|x_2 - x_0\| &\leq \|x_2 - x_1\| + \|x_1 - x_0\| \\
&\leq (1 + I_{h_0}\beta\omega_0(R)f(R)) \|x_1 - x_0\| \\
&\leq (1 + I_{h_0}\beta\omega_0(R)f(R)) \eta \\
&< R.
\end{aligned} \tag{4.39}$$

Second, it is possible to go on with the process because $x_2 \in B(x_0, R) \subset \Omega$. So, there exists Γ_2 and

$$\|\Gamma_2 F'(x_0)\| \leq f(R).$$

In addition, as in the previous step and taking into account that $x_1 + \tau(x_2 - x_1) \in B(x_0, R)$ with $\tau \in [0, 1]$, we have

$$\begin{aligned}
\|F(x_2)\| &= \left\| \int_{x_1}^{x_2} (F'(x) - F'(x_1)) \, dx \right\| \\
&= \left\| \int_0^1 (F'(x_1 + \tau(x_2 - x_1)) \pm F'(x_0) - F'(x_1))(x_2 - x_1) \, d\tau \right\| \\
&\leq 2\beta\omega_0(R)\|x_2 - x_1\|.
\end{aligned}$$

Thus,

$$\|x_3 - x_2\| \leq \|\Gamma_2 F'(x_0)\| \|\Gamma_0\| \|F(x_2)\| \leq 2\beta\omega_0(R)f(R)\|x_2 - x_1\|.$$

Moreover, by applying Lemma 4.26, (4.38) and (4.39), it follows

$$\begin{aligned}
\|x_3 - x_0\| &\leq \|x_3 - x_2\| + \|x_2 - x_0\| \\
&\leq 2\beta\omega_0(R)f(R)\|x_2 - x_1\| + (1 + I_{h_0}\beta\omega_0(R)f(R)) \eta \\
&\leq \left(1 + I_h\beta\omega_0(R)f(R) + 2I_{h_0}(\beta\omega_0(R)f(R))^2\right) \eta \\
&< R.
\end{aligned}$$

After that, we suppose that (i_n)-(ii_n)-(iii_n)-(iv_n) are true for $n = 1, 2, \ldots, j-1$ and prove items (i_j)-(ii_j)-(iii_j)-(iv_j). As $x_j \in B(x_0, R) \subset \Omega$, then there exists Γ_j and

$$\|\Gamma_j F'(x_0)\| \leq f(R).$$

Next, from Taylor's series, we have

$$\|\Gamma_0 F(x_j)\| \leq 2\beta\omega_0(R)\|x_j - x_{j-1}\|. \tag{4.40}$$

Taking now into account

$$\begin{aligned}
\|x_j - x_{j-1}\| &\leq \|\Gamma_{j-1}F'(x_0)\| \|\Gamma_0\| \|F(x_{j-1})\| \tag{4.41} \\
&\leq 2\beta\omega_0(R)f(R)\|x_{j-1} - x_{j-2}\| \\
&\leq \cdots \leq (2\beta\omega_0(R)f(R))^{j-2}\|x_2 - x_1\| \\
&\leq (2\beta\omega_0(R)f(R))^{j-2}I_{h_0}\beta\omega_0(R)f(R)\eta \\
&= 2^{j-2}I_{h_0}(\beta\omega_0(R)f(R))^{j-1}\eta,
\end{aligned}$$

(4.40) and $\|x_j - x_0\| < R$, it follows

$$\|x_{j+1} - x_i\| \leq \|\Gamma_j F'(x_0)\|\|\Gamma_0\|\|F(x_j)\| \leq 2\beta\omega_0(R)f(R)\|x_j - x_{j-1}\|.$$

Then, from (4.41) and the previous lemma, we have

$$
\begin{aligned}
\|x_{j+1} - x_0\| &\leq \|x_{j+1} - x_j\| + \|x_j - x_0\| \\
&\leq \left(1 + I_{h_0}\sum_{k=0}^{j-2} 2^k \left(\beta\omega_0(R)f(R)\right)^{k+1}\right)\eta + 2\beta\omega_0(R)f(R)\|x_j - x_{j-1}\| \\
&\leq \left(1 + I_{h_0}\sum_{k=0}^{j-1} 2^k \left(\beta\omega_0(R)f(R)\right)^{k+1}\right)\eta \\
&< R.
\end{aligned}
$$

Finally, as a consequence, items (i_n)-(ii_n)-(iii_n)-(iv_n) are true for all positive integers n by mathematical induction and the proof is complete. ∎

We are now ready to prove the semilocal convergence of Newton's method under conditions (R1)-(R2)-(R3)-(R4).

Theorem 4.28. *Let $F : \Omega \subseteq X \longrightarrow Y$ be a continuously Fréchet differentiable operator defined on a non-empty open convex domain Ω of a Banach space X with values in a Banach space Y. Suppose that (R1)-(R2)-(R3)-(R4) are satisfied. Then, Newton's sequence $\{x_n\}$ converges to a solution x^* of $F(x) = 0$ starting at x_0. Moreover, $x_n, x^* \in \overline{B(x_0, R)}$ and x^* is unique in $\overline{B(x_0, R)}$.*

Proof. From Lemma 4.27, (4.40) and (4.41), it follows that $\{x_n\}$ is a Cauchy sequence, since

$$
\begin{aligned}
\|x_{n+m} - x_n\| &\leq \|x_{n+m} - x_{n+m-1}\| + \|x_{n+m-1} - x_{n+m-2}\| + \cdots + \|x_{n+1} - x_n\| \\
&\leq \|\Gamma_{n+m-1} F'(x_0)\|\|\Gamma_0\|\|F(x_{n+m-1})\| + \cdots + \|\Gamma_n F'(x_0)\|\|\Gamma_0\|\|F(x_n)\| \\
&\leq f(R)\sum_{j=n}^{n+m-1}\|\Gamma_0\|\|F(x_j)\| \\
&\leq 2\beta\omega_0(R)f(R)\sum_{j=n}^{n+m-1}(2\beta\omega_0(R)f(R))^{j-2}\|x_2 - x_1\| \\
&\leq \left(\sum_{j=n}^{n+m-1}2^{j-1}(\beta\omega_0(R)f(R))^j\right)I_{h_0}\eta \\
&= 2^{n-1}(\beta\omega_0(R)f(R))^n\left(\sum_{j=0}^{m-1}(2\beta\omega_0(R)f(R))^j\right)I_{h_0}\eta \\
&= \frac{1 - (2\beta\omega_0(R)f(R))^{m-1}}{2(1 - 2\beta\omega_0(R)f(R))}(2\beta\omega_0(R)f(R))^n I_{h_0}\eta.
\end{aligned}
$$

Thus, $\{x_n\}$ converges, since $2\beta\omega_0(R)f(R) < 1$. As a consequence, if $\lim x_n = x^*$, then, by letting $j \to +\infty$ in (4.40), we obtain $F(x^*) = 0$ by the continuity of F.

To prove the uniqueness of the solution x^*, we assume that y^* is another solution of $F(x) = 0$ in $\overline{B(x_0, R)}$. Then, from the approximation

$$0 = F(y^*) - F(x^*) = \int_{x^*}^{y^*} F'(x)dx = \int_0^1 F'(x^* + \tau(y^* - x^*))\, d\tau(y^* - x^*),$$

it follows that $x^* = y^*$, provided that the operator $\int_0^1 F'(x^* + \tau(y^* - x^*))\, d\tau$ is invertible. For this, we prove equivalently that there exists the operator J^{-1}, where $J = \Gamma_0 \int_0^1 F'(x^* + \tau(y^* - x^*))\, d\tau$. Indeed, as

$$\|I - J\| = \left\| \int_0^1 \Gamma_0[F'(y^* + \tau(x^* - y^*)) - F'(x_0)]\, d\tau \right\|$$
$$\leq \int_0^1 \beta\omega_0(\|y^* + \tau(x^* - y^*) - x_0\|)\, d\tau$$
$$\leq \beta\omega_0(R)$$
$$< 1,$$

the operator J^{-1} exists by the Banach lemma on invertible operators and the proof is complete. ∎

Finally, we consider an example of nonlinear integral in which we cannot apply the Newton-Kantorovich theorem, but so Theorem 4.28, since the first Fréchet derivative of the operator involved is center Lipschitz in a point of the domain.

Example 4.29. We consider the particular case of integral equation (1.36)-(4.36) given by

$$x(s) = \frac{s}{5} + \frac{1}{2} \int_0^1 G(s, t)x(t)^7\, dt, \quad s \in [0, 1]. \tag{4.42}$$

For the Newton-Kantorovich theorem can be applied, we have previously to locate a solution of integral equation (4.42) in order to obtain the value M and be able to apply the theorem. So, taking into account that a solution $x^*(s)$ of (4.42) in $\mathcal{C}([0, 1])$ must satisfy

$$\|x^*\| - \frac{1}{5} - \frac{1}{16}\|x^*\|^7 \leq 0,$$

since $S = \frac{1}{8}$, it follows that $\|x^*\| \leq \rho_1 = 0.2000\ldots$ or $\|x^*\| \geq \rho_2 = 1.5513\ldots$, where ρ_1 and ρ_2 are the two real positive zeros of the equation deduced from the last expression and given by $\chi(t) = \frac{t^7}{16} - t + \frac{1}{5} = 0$. Thus, from the Newton-Kantorovich theorem, we can only approximate one solution $x^*(s)$ by Newton's method, that which satisfies $\|x^*\| \in [0, \rho_1]$, since we can consider $\Omega = B(0, \rho)$, with $\rho \in (\rho_1, \rho_2)$, where $\mathcal{F}'(x)$ is Lipschitz continuous for the operator $\mathcal{F}'$ corresponding to (4.43) and given by (4.19),

$$[\mathcal{F}'(x)y](s) = y(s) - \frac{7}{2} \int_0^1 G(s, t)x(t)^6 y(t)\, dt,$$

and takes $x_0 \in B(0, \rho)$ as starting point.

Then, we consider $\Omega = B(0, 1)$ and choose the starting point $x_0(s) = 0$. After that, we obtain $\beta = 1$ and $\eta = \frac{1}{5}$, so that the condition of Newton-Kantorovich, $M\beta\eta \leq \frac{1}{2}$, is not satisfied, since $M\beta\eta = 0.525 > \frac{1}{2}$. Therefore, according to the Newton-Kantorovich theorem, we cannot apply Newton's method for approximating a solution of (4.42). However, we can guarantee the convergence of Newton's method from Theorem 4.28, since

$$\omega_0(z) = \frac{7}{16}z, \quad h_0(t) = t, \quad g(t) = \frac{1}{160}(32 - 195t + 210t^2),$$

$g(z)$ has two positive real zeros and the smallest one, $R = 0.2129\ldots$, satisfies the conditions $\beta\omega_0(R) < \frac{1}{3}$, since $\omega_0(R) = 0.0931\ldots$, and $B(x_0, R) = B(0, 0.2129\ldots) \subset B(0, 1) = \Omega$. So, we can now apply Newton's method for approximating a solution of equation (4.42).

4.3.2   Particular case: operators with center Lipschitz first-derivative

Now, we comment the main particular case of operators with center ω-Lipschitz first derivate that we already know: the center Lipschitz case.

If $\omega_0(z) = M_0z$ and $h_0(t) = t$ in (R2), F' is center Lipschitz continuous in Ω and conditions (R1)-(R2)-(R3)-(R4) are reduced to the following ones:

(S1) There exists $\Gamma_0 = [F'(x_0)]^{-1} \in \mathcal{L}(Y, X)$, for some $x_0 \in \Omega$, with $\|\Gamma_0\| \leq \beta$ and $\|\Gamma_0 F(x_0)\| \leq \eta$.

(S2) There exists a constant $M_0 \geq 0$ such that $\|F'(x) - F'(x_0)\| \leq M_0\|x - x_0\|$, for $x \in \Omega$,

(S3) $\ell_0 = M_0\beta\eta \leq \frac{14-4\sqrt{6}}{25} = 0.1680816\ldots$

(S4) $B(x_0, R) \subset \Omega$, where $R = \frac{2+5\ell_0 - \sqrt{(2+5\ell_0)^2 - 48\ell_0}}{12M_0\beta}$.

As a consequence, the semilocal convergence of Newton's method is now guaranteed from the next result given in [39].

Corollary 4.30. *Let $F : \Omega \subseteq X \longrightarrow Y$ be a continuously Fréchet differentiable operator defined on a non-empty open convex domain Ω of a Banach space X with values in a Banach space Y. Suppose that (S1)-(S2)-(S3)-(S4) are satisfied. Then, Newton's sequence $\{x_n\}$ converges to a solution x^* of $F(x) = 0$ starting at x_0. Moreover, $x_n, x^* \in \overline{B(x_0, R)}$ and x^* is unique in $\overline{B(x_0, R)}$.*

Remark 4.31. Observe that, from condition (S2) with $M_0 \leq M$, it follows that condition $M_0\beta\eta \leq \zeta$ with $\zeta < \frac{1}{2}$ can be satisfied (see condition (S3)). Moreover, as $M_0 \leq M$, condition $M\beta\eta \leq \frac{1}{2}$ can be not satisfied, so that we can find new starting points x_0 for Newton's method.

In the next example, we present a nonlinear integral equation where the first Fréchet derivative of the operator involved can be considered Hölder continuous in the domain or center Lipschitz at a point of the domain and see that, in this case, the second situation is more favourable than the first.

Example 4.32. We now consider the particular case of integral equation (1.36)-(4.17) given by

$$x(s) = 1 + 2\int_0^1 G(s, t)x(t)^{\frac{4}{3}}\, dt, \quad s \in [0, 1]. \tag{4.43}$$

In this case, the operator $\mathcal{F}'$ corresponding to (4.43) and given by (4.19) is

$$[\mathcal{F}'(x)y](s) = y(s) - \frac{8}{3} \int_0^1 G(s,t)x(t)^{\frac{1}{3}}y(t)\, dt,$$

which is Hölder continuous in $\Omega = \mathcal{C}([0,1])$, where Hölder's constants are $K = \frac{1}{3}$ and $p = \frac{1}{3}$, so that we can use Theorem 4.23 to guarantee the semilocal convergence of Newton's method to a solution of equation (4.43).

If we choose $x_0(s) = 1$, then $\beta = \frac{3}{2}$, $\eta = \frac{3}{8}$, $b_0 = K\beta\eta^p = 0.3605\ldots$ and the condition $b_0 \le \xi$, where ξ is the smallest positive real zero of function (4.30), $\phi(t;p) = \frac{2^{2/3}(1-t)^{4/3}}{\sqrt[3]{3}} - \sqrt[3]{t} = 0$, is not satisfied, since $\xi = 0.3071\ldots$ However, we can guarantee the convergence of Newton's method from Theorem 4.28, since

$$\omega_0(z) = \frac{z}{3}, \quad h_0(t) = t, \quad g(t) = \frac{1}{16}(6 - 21t + 16t^2),$$

$g(z)$ has two positive zeros and the smallest one, $R = 0.4203\ldots$, satisfies the conditions $\beta\omega_0(R) < \frac{1}{3}$, since $\omega_0(R) = 0.1401\ldots$, and $B(x_0, R) = B(1, 0.4203\ldots) \subset \Omega$. So, we can now apply Newton's method for approximating a numerical solution of equation (4.43) by using first a process of discretization, as we have done in Example 4.20.

Finally, we note that other particular case of operators with center ω-Lipschitz first derivative is the center Hölder case, which is given when $\omega_0(z) = K_0 z^p$ and $h_0(t) = t^p$ with $p \in [0,1]$ in (R2).

4.3.3 Application

In this section, we provide some results of existence and uniqueness of solution for nonlinear integral equations of Hammerstein-type (1.36)-(4.17) with $p = \frac{1}{n}$, $n \in \mathbb{N}$, $q \in \mathbb{N}$, $q \ge 3$ and $\lambda_1, \lambda_2 \in \mathbb{R}$; i.e.:

$$x(s) = u(s) + \int_a^b \mathcal{K}(s,t)\left(\lambda_1\, x(t)^{1+\frac{1}{n}} + \lambda_2\, x(t)^q\right) dt, \quad s \in [a,b], \tag{4.44}$$

where u is a continuous function such that $u(s) > 0$ and the kernel $\mathcal{K}(s,t)$ is continuous and nonnegative in $[a,b] \times [a,b]$.

Existence and uniqueness of solution

Observe that solving the equation (4.44) is equivalent to solving $\mathcal{F}(x) = 0$, where $\mathcal{F} : \Omega \subseteq \mathcal{C}([a,b]) \longrightarrow \mathcal{C}([a,b])$ and

$$[\mathcal{F}(x)](s) = x(s) - u(s) - \int_a^b \mathcal{K}(s,t)\left(\lambda_1 x(t)^{1+\frac{1}{n}} + \lambda_2 x(t)^q\right) dt, \tag{4.45}$$

where $n \in \mathbb{N}$, $q \in \mathbb{N}$, $q \ge 3$ and $\lambda_1, \lambda_2 \in \mathbb{R}$. Now, we choose the domain $\Omega \subseteq \mathcal{C}([a,b])$. For this, we consider

$$\Omega = \{x \in \mathcal{C}([a,b]) : x(s) > 0, \ \|x(s)\| < \rho, \ s \in [a,b]\} \text{ if } n \text{ is even}, \tag{4.46}$$
$$\Omega = \{x \in \mathcal{C}([a,b]) : \|x(s)\| < \rho, \ s \in [a,b]\} \text{ if } n \text{ is odd},$$

since if equation (4.44) has a solution $x^*(s)$, this solution must be located previously in a ball $B(0, \rho)$.

According to what we have seen in the introduction of Section 4.3 for integral equations (4.34) and (4.36), it is clear that $\|\mathcal{F}'(x) - \mathcal{F}'(x_0)\| \leq M_0 \|x - x_0\|$, where

$$M_0 = S \left(\frac{|\lambda_1| \left(1 + \frac{1}{n}\right)}{d^{\frac{n-1}{n}}} + |\lambda_2| q \sum_{i=0}^{q-2} \|x_0\|^i \rho^{q-2-i} \right), \tag{4.47}$$

$S = \left\| \int_a^b \mathcal{K}(s,t) \, dt \right\|$, $x_0(s) = u(s)$ and $d = \min_{s \in [a,b]} u(s) > 0$. As a consequence, the operator $\mathcal{F}'$ is center Lipschitz in Ω.

On the other hand, following what has been done to obtain the parameters β and η for operator (4.18), we have

$$\|[\mathcal{F}'(x_0)]^{-1}\| = \|[\mathcal{F}'(u)]^{-1}\| \leq \frac{1}{1 - S\left(|\lambda_1| \left(1 + \frac{1}{n}\right) \left\|u^{\frac{1}{n}}\right\| + |\lambda_2| q \|u^{q-1}\|\right)} = \beta, \tag{4.48}$$

$$\|[\mathcal{F}'(x_0)]^{-1}\mathcal{F}(x_0)\| = \|[\mathcal{F}'(u)]^{-1}\mathcal{F}(u)\| \leq \frac{S\left(|\lambda_1| \left\|u^{1+\frac{1}{n}}\right\| + |\lambda_2| \|u^q\|\right)}{1 - S\left(|\lambda_1| \left(1 + \frac{1}{n}\right) \left\|u^{\frac{1}{n}}\right\| + |\lambda_2| q \|u^{q-1}\|\right)} = \eta, \tag{4.49}$$

provided that $S\left(|\lambda_1| \left(1 + \frac{1}{n}\right) \left\|u^{\frac{1}{n}}\right\| + |\lambda_2| q \|u^{q-1}\|\right) < 1$.

Once the parameters β and η are calculated, we can already establish the following result on the existence of solution of equation (4.44) from Corollary 4.30.

Theorem 4.33. *Let $\mathcal{F}$ be the operator defined in (4.45) with Ω given in (4.46) and $u \in \Omega$ such that $S\left(|\lambda_1| \left(1 + \frac{1}{n}\right) \left\|u^{\frac{1}{n}}\right\| + |\lambda_2| q \|u^{q-1}\|\right) < 1$. If $\ell_0 = M_0 \beta \eta \leq \frac{14 - 4\sqrt{6}}{25} = 0.1680816\ldots$ with M_0 given in (4.47), β in (4.48), η in (4.49), and $B(u, R) \subset \Omega$, where $R = \frac{2 + 5\ell_0 - \sqrt{(2+5\ell_0)^2 - 48\ell_0}}{12 M_0 \beta}$, then a unique solution of (4.44) exists in $\overline{B(u, R)}$.*

Note also that the bound given for $\|\mathcal{F}(x_0)\|$ can be improved when the kernel $\mathcal{K}(s,t)$ is fixed.

Example 4.34. We consider the following particular integral equation of type (4.44):

$$x(s) = 1 + \frac{1}{2} \int_0^1 G(s,t) \left(x(t)^{\frac{4}{3}} + x(t)^3\right) dt \quad s \in [0, 1]. \tag{4.50}$$

To apply Theorem 4.33, we have previously to locate a solution of integral equation (4.50) in order to obtain the value M_0 and be able to apply the theorem. So, taking into account that a solution $x^*(s)$ of (4.50) in $\mathcal{C}([a,b])$ must satisfy

$$\|x^*\| - 1 - \frac{1}{16} \left(\|x^*\|^{\frac{4}{3}} + \|x^*\|^3\right) \leq 0,$$

since $S = \frac{1}{8}$, it follows that $\|x^*\| \leq \rho_1 = 1.1809\ldots$ or $\|x^*\| \geq \rho_2 = 3.0503\ldots$, where ρ_1 and ρ_2 are the two real positive zeros of the function deduced from the last expression and given by $\chi(t) = \frac{1}{16}\left(t^{\frac{4}{3}} + t^3\right) - t + 1$. Thus, from Theorem 4.33, we can only guarantee the existence of a solution $x^*(s)$ of integral equation (4.50) if $\|x^*\| \in [0, \rho_1]$, since we can consider $\Omega = B(0, \rho)$, with $\rho \in (\rho_1, \rho_2)$, where $F'(x)$ is center Lipschitz continuous.

Taking into account $x_0(s) = u(s) = 1$ and choosing $\rho = 2$, we have

$$n = \frac{1}{3}, \quad q = 3, \quad \lambda_1 = \lambda_2 = \frac{1}{2}, \quad d = 1, \quad \beta = \frac{48}{35}, \quad \eta = \frac{6}{35}, \quad M_0 = \frac{31}{48}, \quad R = 0.2959\ldots$$

Thus, as $S\left(|\lambda_1|\left(1 + \frac{1}{n}\right)\left\|x_0^{\frac{1}{n}}\right\| + |\lambda_2|q\|x_0^{q-1}\|\right) = 0.2708\ldots < 1$, $\ell_0 = M_0\beta\eta = 0.1518\ldots \le$
$\frac{14-4\sqrt{6}}{25} = 0.1680816\ldots$ and $B(u, R) = B(1, 0.2959\ldots) \subset B(0, 2) = \Omega$, the hypotheses of Theorem 4.33 are satisfied. As a consequence, (4.50) has a unique solution $x^*(s)$ in the region $\{\nu \in \mathcal{C}([0, 1]) : \|\nu - 1\| \le 0.2959\ldots\}$.

Bibliography

[1] A. A. Andronow and C. E. Chaikin, *Theory of oscillations*, Princenton University Press, New Jersey, 1949.

[2] I. K. Argyros, *Polynomial operator equations in abstract spaces and applications*, CRC Press, Boca Raton, 1998.

[3] I. K. Argyros, *On the convergence of Newton's method for polynomial equations and applications in radiative transfer*, Monatsh. Math., 127, 4 (1999) 265–276.

[4] I. K. Argyros, *An improved convergence analysis and applications for Newton-like methods in Banach space*, Numer. Funct. Anal. Optim., 24, 7-8 (2003) 653–572.

[5] I. K. Argyros, *On the Newton-Kantorovich hypothesis for solving equations*, J. Comput. Appl. Math., 169, 2 (2004) 315–332.

[6] K. E. Atkinson, *The numerical solution of a nonlinear boundary integral equation on smooth surfaces*, IMA J. Numer. Anal., 14 (1994) 461–483.

[7] K. E. Atkinson, *The numerical solution of integral equations of the second kind*, Cambridge University Press, Cambridge, 1997.

[8] A. T. Bharucha-Reid and R. Kannan, *Newton's method for random operator equations*, J. Nonlinear Anal., 4 (1980) 231–240.

[9] J. Banás, C. J. Rocha Martin and K. Sadarangani, *On solutions of a quadratic integral equation of Hammerstein type*, Math. Comput. Modelling, 43 (2006) 97–104.

[10] D. D. Bruns and J. E. Bailey, *Nonlinear feedback control for operating a nonisothermal CSTR near an unstable steady state*, Chem. Eng. Sci., 32 (1977) 257–264.

[11] L. Collatz, *Functional analysis and numerical mathematics*, Academic Press, New York, 1966.

[12] C. Corduneanu, *Integral Equations and Applications*, Cambridge University Press, Cambridge, 1991.

[13] H. T. Davis, *Introduction to Nonlinear Differential and Integral Equations*, Dover Pub., New York, 1962.

[14] K. Deimling, *Nonlinear functional analysis*, Springer-Verlag, Berlin, 1985.

© Springer International Publishing AG 2017

J.A. Ezquerro Fernández, M.Á. Hernández Verón, *Newton's Method: an Updated Approach of Kantorovich's Theory*, Frontiers in Mathematics, DOI 10.1007/978-3-319-55976-6

[15] J. E. Dennis, *On the Kantorovich hypotheses for Newton's method*, SIAM J. Numer. Anal., 6 (1969) 493–507.

[16] J. E. Dennis and R. B. Schnabel, *Numerical methods for unconstrained optimization and nonlinear equations*, SIAM, Philadelphia, 1996.

[17] P. Deulfhard and G. Heindl, *Affine invariant convergence theorems for Newton's method and extensions to related methods*, SIAM J. Numer. Anal., 16 (1979) 1–10.

[18] P. Deulfhard, *Newton methods for nonlinear problems. Affine invariant and adaptative algorithms*, Springer, Berlin, 2004.

[19] P. Deulfhard, *A short history of Newton's method*, Documenta Math., 25 (2012) 25–30.

[20] J. A. Ezquerro and M. A. Hernández, *On an application of Newton's method to nonlinear operators with w-conditioned second derivative*, BIT, 42, 3 (2002) 519–530.

[21] J. A. Ezquerro and M. A. Hernández, *Generalized differentiability conditions for Newton's method*, IMA J. Numer. Anal., 22, 2 (2002) 187–205.

[22] J. A. Ezquerro and M. A. Hernández, *On the R-order of convergence of Newton's method under mild differentiability conditions*, J. Comput. Appl. Math., 197, 1 (2006) 53–61.

[23] J. A. Ezquerro and M. A. Hernández, *Halley's method for operators with unbounded second derivative*, Appl. Numer. Math., 57, 3 (2007) 354–360.

[24] J. A. Ezquerro, J. M. Gutiérrez, M. A. Hernández, N. Romero, M. J. Rubio, *The Newton method: from Newton to Kantorovich* (Spanish), Gac. R. Soc. Mat. Esp., 13, 1 (2010) 53–76.

[25] J. A. Ezquerro, D. González and M. A. Hernández *Majorizing sequences for Newton's method from initial value problem*, J. Comput. Appl. Math., 236 (2012) 2246–2258.

[26] J. A. Ezquerro, D. González and M. A. Hernández, *A general semilocal convergence result for Newton's method under centered conditions for the second derivative*, ESAIM-Math. Model. Numer. Anal., 47, 1 (2013) 149–167.

[27] J. A. Ezquerro, D. González and M. A. Hernández, *A modification of the classic conditions of Newton-Kantorovich for Newton's method*, Math. Comput. Modelling, 57 (2013) 584–594.

[28] J. A. Ezquerro, D. González and M. A. Hernández-Verón, *A semilocal convergence result for Newton's method under generalized conditions of Kantorovich*, J. Complexity, 30 (2014) 309–324.

[29] J. A. Ezquerro and M. A. Hernández-Verón, *Center conditions on high order derivatives in the semilocal convergence of Newton's method*, J. Complexity, 31 (2015) 277–292.

[30] J. A. Ezquerro, M. A. Hernández-Verón and A. I. Velasco *An analysis of the semilocal convergence for secant-like methods*, Appl. Math. Comp., 266 (2015) 883–892.

[31] F. Faraci and V. Moroz, *Solutions of Hammerstein integral equations via a variational principle*, J. Integral Equations Appl., 15, 4 (2003) 385–402.

[32] I. Fenyö, *Über die Lösung der im Banachsen Raume definierten nichtlinearen Gleichungen*, Acta Math. Hungar., 5 (1954) 85–93.

[33] A. Galántai, *The theory of Newton's method*, J. Comput. Appl. Math., 124 (2000) 25–44.

[34] M. Ganesh and M. C. Joshi, *Numerical solvability of Hammerstein integral equations of mixed type*, IMA J. Numer. Anal. 11 (1991) 21–31.

[35] H. H. Goldstine, *A history of Numerical Analysis from the 16th through the 19th Century*, Springer, Berlin, 1977.

[36] W. B. Gragg and R. A. Tapia, *Optimal error bounds for the Newton-Kantorovich theorem*, SIAM J. Numer. Anal., 11, 1 (1974) 10–13.

[37] J. M. Gutiérrez, *A new semilocal convergence theorem for Newton's method*, J. Comput. Appl. Math., 79 (1997) 131–145.

[38] J. M. Gutiérrez and M. A. Hernández, *Majorizing sequences for Newton's method*, Tamkang J. Math., 29, 3 (1998) 199–202.

[39] J. M. Gutiérrez and M. A. Hernández, *Newton's method under weak Kantorovich conditions*, IMA J. Numer. Anal., 20 (2000) 521–532.

[40] J. M. Gutiérrez and M. A. Hernández, *Newton's method under different Lipschitz conditions*, (Numerical analysis and its applications) Lectures Notes in Comput. Sci., 1988 (2000) 368–376.

[41] M. A. Hernández and M. A. Salanova, *Indices of convexity and concavity. Application to Halley method*, Appl. Math. Comput., 103 (1999) 27–49.

[42] M. A. Hernández, *Relaxing convergence conditions for Newton's method*, J. Math. Anal. Appl., 249 (2000) 463–475.

[43] M. A. Hernández, *The Newton method for operators with Hölder continuous first derivative*, J. Optim. Theory Appl., 109 (2001) 631–648.

[44] S. Hu, M. Khavanin and W. Zhuang, *Integral equations arising in the kinetic theory of gases*, Appl. Anal., 34 (1989) 261–266.

[45] Huang Zhengda, *A note on the Kantorovich theorem for Newton method*, J. Comput. Appl. Math., 47 (1993) 211–217.

[46] L. V. Kantorovich, *The method of successive approximations for funcitonal equations*, Acta Math., 71 (1939) 63–97.

[47] L. V. Kantorovich, *On Newton's method for functional equations* (Russian), Dokl. Akad. Nauk. SSSR, 59 (1948) 1237–1240.

[48] L. V. Kantorovich, *Functional analysis and applied mathematics* (Russian), Uspekhi Mat. Nauk, 3 (1948) 89–185.

[49] L. V. Kantorovich, *On Newton's method* (Russian), Trudy Math. Inst. Steklov, 28 (1949) 104–144.

[50] L. V. Kantorovich, *The majorant principle and Newton's method* (Russian), Dokl. Akad. Nauk SSSR, 76 (1951) 17–20.

[51] L. V. Kantorovich, *Some further applications of principle of majorants* (Russian), Dokl. Akad. Nauk SSSR, 80 (1951) 849–852.

[52] L. V. Kantorovich, *On approximate solution of functional equations* (Russian), Uspekhi Mat. Nauk, 11 (1956) 99–116.

[53] L. V. Kantorovich, *Some further applications of Newton's method* (Russian), Vest. LGU, Ser. Math. Mech., 7 (1957) 68–103.

[54] L. V. Kantorovich and G. P. Akilov, *Functional analysis in normed spaces* (Russian), Fizmatgiz, Moscow, 1959; translated by D. E. Brown and A. P. Robertson, Pergamon Press, Oxford, 1964.

[55] L. V. Kantorovich and G. P. Akilov, *Functional analysis*, Pergamon Press, New York, 1982.

[56] H. B. Keller, *Numerical methods for two-point boundary value problems*, Dover Pub., New York, 1992.

[57] N. Kollerstrom, *Thomas Simpson and Newton's method of approximation: an enduring myth.* British Journal for History of Science, 25 (1992) 347–354.

[58] M. A. Krasnoselski, G. M. Vainikko, P. P. Zabreiko, Ya. B. Rutitski and V. Ja. Stetsenko, *Approximate Solution of Operator Equations* (Russian), Nauka, Moscow, 1969; translation in Wolters-Noordhoff Publishing, Groningen, 1972; & Akademie-Verlag, Berlin, 1973.

[59] J. H. Mathews, *Bibliography for Newton's method*,
http://mathfaculty.fullerton.edu/mathews/n2003/newtonsmethod/
Newton'sMethodBib/Links/Newton'sMethodBib_lnk_2.html

[60] G. J. Miel, *The Kantorovich theorem with optimal error bounds*, Amer. Math. Monthly, 86 (1979) 212–215.

[61] I. P. Mysovskikh, *On convergence of Newton's method* (Russian), Trudy Mat. Inst. Steklov, 28 (1949) 145–147.

[62] I. P. Mysovskikh, *On convergence of L. V. Kantorovich's method for functional equations and its applications* (Russian), Dokl. Akad. Nauk SSSR, 70 (1950) 565–568.

[63] J. M. Ortega, *The Newton-Kantorovich theorem*, Amer. Math. Monthly, 75 (1968) 658–660.

[64] J. M. Ortega and W. C. Rheinboldt, *Iterative solution of nonlinear equations in several variables*, Academic Press, New York, 1970.

[65] A. M. Ostrowski, *Solution of equations and systems of equations*, Academic Press, New York, 1966.

[66] B. T. Polyak, *Newton-Kantorovich method and its global convergence* (English, Russian summary), Zap. Nauchn. Sem. S.-Peterburg. Otdel. Mat. Inst. Steklov. (POMI) 312 (2004), Teor. Predst. Din. Sist. Komb. i Algoritm. Metody. 11, 256–274, 316; translation in J. Math. Sci. 133, 4 (2006) 1513–1523.

[67] A. D. Polyanin and A. V. Manzhirov, *Handbook of integral equations*, CRC Press, Boca Raton, 1998.

[68] D. Porter and D. Stirling, *Integral equations: a practical treatment, from spectral theory to applications*, Cambridge University Press, Cambridge, 1990.

[69] F. A. Potra and V. Pták, *Nondiscrete induction and iterative methods*, Pitman Publishing Limited, London, 1984.

[70] J. Rashidinia and M. Zarebnia, *New approach for numerical solution of Hammerstein integral equations*, Appl. Math. Comput., 185 (2007) 147–154.

[71] L. B. Rall, *A note on the convergence of Newton?s method*, SIAM J. Numer. Anal., 11 (1974) 34–36.

[72] L. B. Rall, *Computational solution of nonlinear operator equations*, Robert E. Krieger Publishing Company, Michigan, 1979.

[73] W. C. Rheinboldt, *A unified convergence theory for a class of iterative processes*, SIAM J. Numer. Anal., 5 (1968) 42–63.

[74] W. C. Rheinboldt, *Methods for solving systems of nonlinear equations*, SIAM, Philadelphia, 1998.

[75] J. Rokne, *Newton's method under mild differentiability conditions with error analysis*, Numer. Math. 18 (1972) 401–412.

[76] E. Schröeder, *Über unendlich viele algotithmen zur aufl¨ösung der Gleichugen*, Math. Ann. 2 (1870) 317–365.

[77] J. J. Stoker, *Nonlinear vibrations*, Interscience-Wiley, New York, 1950.

[78] R. A. Tapia, *The Kantorovich theorem for Newton's method*, Amer. Math. Monthly, 78 (1971) 389–392.

[79] B. A. Vertgeim, *On the conditions of Newton's approximation method* (Russian), Dokl. Akad. Nauk SSSR, 110 (1956) 719–722.

[80] T. Yamamoto, *A method for finding sharp error bounds for Newton's method under the Kantorovich assumptions*, Numer. Math., 49 (1986) 203–220.

[81] T. Yamamoto, *Convergence theorem for Newton-like methods in Banach spaces*, Numer. Math., 51 (1987) 545–557.

[82] T. Yamamoto, *On the method of tangent hyperbolas in Banach spaces*, J. Comput. Appl. Math., 21, 1 (1988) 75–86.

[83] T. Yamamoto, *Historical developments in convergence analysis for Newton's and Newton-like methods*, J. Comput. Appl. Math., 124 (2000) 1–23.

[84] L. Yau and A. Ben-Israel, *The Newton and Halley methods for complex roots*, Amer. Math. Monthly, 105 (1998) 806–818.

[85] T. J. Ypma, *Historical development of the Newton-Raphson method*, SIAM Review, 37 (1995) 531–551.

[86] Z. Zhang, *A note on weaker convergence conditions for Newton iteration* (Chinese), J. Zhejiang Univ. Sci. Ed., 30, 2 (2003) 133–135, 144.